HIGH-k GATE DIELECTRIC MATERIALS

Applications with Advanced Metal Oxide Semiconductor Field Effect Transistors (MOSFETs)

HIGH-k GATE DIELECTRIC MATERIALS

Applications with Advanced Metal Oxide Semiconductor Field Effect Transistors (MOSFETs)

Edited by

Niladri Pratap Maity, PhD

Reshmi Maity, PhD

Srimanta Baishya, PhD

AAP APPLE ACADEMIC PRESS

First edition published [2021]

Apple Academic Press Inc.
1265 Goldenrod Circle, NE,
Palm Bay, FL 32905 USA

4164 Lakeshore Road, Burlington,
ON, L7L 1A4 Canada

CRC Press
6000 Broken Sound Parkway NW,
Suite 300, Boca Raton, FL 33487-2742 USA

2 Park Square, Milton Park,
Abingdon, Oxon, OX14 4RN UK

First issued in paperback 2021

Apple Academic Press exclusively co-publishes with CRC Press, an imprint of Taylor & Francis Group, LLC

Library and Archives Canada Cataloguing in Publication

Title: High-k gate dielectric materials : applications with advanced metal oxide semiconductor field effect transistors (MOSFETs) / edited by Niladri Pratap Maity, PhD, Reshmi Maity, PhD, Srimanta Baishya, PhD.

Names: Maity, Niladri Pratap, editor. | Maity, Reshmi, editor. | Baishya, Srimanta, editor.

Description: Includes bibliographical references and index.

Identifiers: Canadiana (print) 20200220306 | Canadiana (ebook) 20200220462 | ISBN 9781771888431 (hardcover) | ISBN 9780429325779 (ebook)

Subjects: LCSH: Gate array circuits—Materials. | LCSH: Metal oxide semiconductors—Materials. | LCSH: Dielectrics.

Classification: LCC TK7871.99.M44 H54 2020 | DDC 621.3815/2—dc23

Library of Congress Cataloging-in-Publication Data

Names: Maity, Niladri Pratap, editor. | Maity, Reshmi, editor. | Baishya, Srimanta, editor.

Title: High-k gate dielectric materials : applications with advanced metal oxide semiconductor field effect transistors (MOSFETs) / edited by Niladri Pratap Maity, PhD, Reshmi Maity, PhD, Srimanta Baishya, PhD.

Description: 1st edition. | Palm Bay, Florida : Apple Academic Press, [2020] | Includes bibliographical references and index. | Summary: "This volume explores and addresses the challenges of high-k gate dielectric materials, one of the major concerns in the evolving semiconductor industry and the International Technology Roadmap for Semiconductors (ITRS). The application of high-k gate dielectric materials is a promising strategy that allows further miniaturization of microelectronic components (or Moore's law). This book presents a broad review of SiO2 materials, including a brief historical note of Moore's law, followed by reliability issues of the SiO2 based MOS transistor. Then it discusses the transition of gate dielectrics with an EOT 1 nm and a selection of high-k materials. A review of the different deposition techniques of different high-k films is also discussed. High-k dielectrics theories (quantum tunneling effects and interface engineering theory) and applications of different novel MOSFET structures, like tunneling FET, are also covered in this book. The volume also looks at the important issues in the future of CMOS technology and presents an analysis of interface charge densities with the high-k material tantalum pentoxide. The issue of CMOS VLSI technology with the high-k gate dielectric materials is covered as is the advanced MOSFET structure, with its working, structure, and modeling. This timely volume addresses the challenges of high-k gate dielectric materials and will prove to be a valuable resource on both the fundamentals and the successful integration of high-k dielectric materials in future IC technology. Key features: Discusses the state-of-the-art in high-k gate dielectric research for MOSFET in the nanoelectronics regime Reviews high-k applications in advanced MOS transistor structures Considers CMOS IC fabrication with high-k gate dielectric materials"-- Provided by publisher.

Identifiers: LCCN 2020015811 (print) | LCCN 2020015812 (ebook) | ISBN 9781771888431 (hardcover) | ISBN 9780429325779 (ebook)

Subjects: LCSH: Dielectrics. | Gate array circuits. | Metal oxide semiconductor field-effect transistors.

Classification: LCC QC585 .H537 2020 (print) | LCC QC585 (ebook) | DDC 621.3815/2840284--dc23

LC record available at https://lccn.loc.gov/2020015811

LC ebook record available at https://lccn.loc.gov/2020015812

ISBN: 978-1-77188-843-1 (hbk)
ISBN: 978-1-77463-885-9 (pbk)
ISBN: 978-0-42932-577-9 (ebk)

About the Editors

Niladri Pratap Maity, PhD, is an Associate Professor in the Department of Electronics and Communication Engineering at Mizoram University (a Central University), India. He is the author of more than 120 journal articles and conference papers. He has been nominated for and selected as a Visiting Scientist by the Department of Science and Technology, Government of India. He is a recipient of several best/excellent paper awards. His research interests include VLSI design, MOS device modeling, and MEMS. Dr. Maity received MSc degree in Electronics from Sambalpur University, India, MTech degree in Electronics Design and Technology from Tezpur University (A Central University), Tezpur, India, and a PhD degree in Electronics and Communication Engineering from the National Institute of Technology, Silchar, India.

Reshmi Maity, PhD, is an Associate Professor in the Department of Electronics and Communication Engineering, Mizoram University (a Central University), Aizawl, India. Prior to that, she was an Assistant Professor at the JIS College of Engineering (West Bengal University of Technology) at Kolkata, India. She is the author of over 100 refereed publications. Her research interests include nanoelectronics and MEMS. She received her BTech and MTech degrees in Radio Physics and Electronics from the University of Calcutta, Kolkata, India. She has completed her PhD degree in Electronics and Communication Engineering at the National Institute of Technology, Silchar, India.

Srimanta Baishya, PhD, is a Professor in the Department of Electronics and Communication Engineering of the National Institute of Technology Silchar, India. Prior to that, he was an Assistant Professor in the Department of Electronics and Telecommunication Engineering of the same college. He was also with the Department of Electrical Engineering as a Lecturer in the then REC Silchar, India. His research interests cover semiconductor devices and circuits, MOS transistor modeling, and MEMS-based energy harvesting. He has published over 90 papers in peer-reviewed journals. Dr. Baishya is a senior member of the IEEE and a Fellow of the Institution of Engineers (India). He received his BE degree from the Assam Engineering College, Guwahat, India; MTech degree from IIT Kanpur, both in Electrical Engineering; and PhD degree in Engineering from Jadavpur University, Kolkata, India.

Contents

Contents

Contributors

Prof. Srimanta Baishya, Professor and Dean (Academic)
Department of Electronics and Communication Engineering, National Institute of Technology, Silchar-788010, India

Prof. V. Madhurima, Professor
Department of Physics, School of Basic and Applied Sciences, Central University of Tamilnadu, Thiruvarur, Tamilnadu, India

Dr. Niladri Pratap Maity, Associate Professor and Head
Department of Electronics and Communication Engineering, Mizoram University (A Central University), Aizawl-796004, India

Dr. Reshmi Maity, Associate Professor
Department of Electronics and Communication Engineering, Mizoram University (A Central University), Aizawl-796004, India

Dr. D. P. Rai, Assistant Professor
Physical Sciences Research Center (PSRC), Department of Physics, Pachhunga University College, Aizawl-796001, India

Prof. P. P. Sahu, Professor and Head
Department of Electronics and Communication Engineering, Tezpur University (A Central University), Tezpur-784028, India

P. Sri Harsha, Research Scholar
Department of Physics, School of Basic and Applied Sciences, Central University of Tamilnadu, Thiruvarur, Tamilnadu, India

K. Venkata Saravanan, Research Scholar
Department of Physics, School of Basic and Applied Sciences, Central University of Tamilnadu, Thiruvarur, Tamilnadu, India

Abbreviations

ALD	atomic layer deposition
APCVD	atmospheric pressure CVD
ATRP	atom transfer radical polymerization
BOPP	biaxially oriented polypropylene
BTO	$BaTiO_3$
BTS	bias temperature stress
BST	barium strontium titanate
CB	conduction band
CBM	conduction band minima
CE	cyanate ester
CMOS	complementary metal-oxide-semiconductor
CuPc	copperphthalocyanine
CVD	chemical vapor deposition
CYELP	cyanoethyl pullulan
DFT	density functional theory
DIBL	drain-induced barrier lowering
DIP	dual in-line packaging
DOS	density of states
DOF	depth of focus
DRAM	dynamic random access memory
ECB	electron conduction band
EHP	electron hole pair
EOT	effective oxide thickness
EOT	equivalent oxide thickness
EAP	electro-active polymer
EVB	electron valence band
FET	field-effect transistor
FHA	full-Heusler alloy
FHD	flame hydrolysis deposition
FIBL	field-induced barrier lowering
FLAPW	full potential linearized augmented plane wave
FM	ferromagnetic
GIDL	gate-induced drain leakage

GGA	generalized gradient approximation
GN	graphene nanosheet
GMR	giant magnetoresistance
HHA	half-Heusler alloy
HMF	half-metallic ferromagnetic
ICs	integrated circuits
LDA	local density approximation
LPCVD	low pressure CVD
LSIs	large-scale integrated circuits
MEMS	micro-electro-mechanical systems
MMA	methyl methacrylate
MOCVD	metal-organic CVD
MOS	metal oxide semiconductor
MRAM	magnetoresistive random-access memory
MtPc	metallophthalocyanine
MWNT	multiwalled carbon nanotube
NM	non-magnetic
OFET	organic field-effect transistor
PANI	polyaniline
PBE	Perdew, Burke, and Ernzerh
PBS	poly-butene-1-sulfone
PDP	power delay product
PECVD	plasma-enhanced CVD
PMMA	poly-methyle-metarylate
PSF	polysulfone
PVDF	polyvinylidene fluoride
RAFT	reversible addition-fragmentation chain transfer
RBA	rigid-band approximation
RF	radio frequency
RIE	reactive ion etching
RIBE	reactive ion beam etching
RTA	rapid thermal annealing
SCEs	short-channel effects
SCL	space charge limited
SIMS	secondary ion mass spectrometry
SS	subthreshold swing
TCA	trichloroethane
TCE	trichloroethylene

TCAD	technology computer-aided design
TM	transition metal
TMR	tunneling magnetoresistance
ULSIs	ultra large-scale integrated circuits
VBM	valence band maxima

Preface

"*There is no reason anyone would want a computer in their home.*" This quote came from the lips of K. H. Olson, Chairman and Founder of Digital Equipment Corporation in 1977. At that period, the first generation microprocessor had only come to commercial market. Nowadays, having at least one personal computer is natural in each household, and small electronic gadgets, such as smartphones, tablets etc., are very common and essential for every person. This terrific growth of the electronics industry is the consequence of several decades of enormous improvement in the Large Scale Integrated Circuits (LSIs) technology. The recent historic progress of electronics and information technology is owed to high speed and smaller devices. This advancement has improved quality of life, which significantly contributed to the development of civilization and is indispensable in modern societies.

The proposed concept of Lilienfeld in 1930 on a Metal Oxide Semiconductor Field Effect Transistor (MOSFET) was realized much later by Kahgn and Attala in 1960 using Si substrate and SiO_2 as the gate insulator. Since then, Si and SiO_2 have become the key materials for electronic devices and circuits. The gate-insulator-substrate that controls the on/off state of the device is the heart of the MOS transistor and is commonly known as MOS capacitor or two-terminal MOS transistor. The MOS structure, first proposed as a voltage–variable capacitor in 1959 by Moll, is the most important component of solid state electronics. Since the reliability and stability of semiconductor devices are intimately related to their surface and interface conditions, an understanding of the physics of surface and interface of the MOS devices is of great importance for exploring the proper device operations.

A MOSFET is usually fabricated by oxidizing a Si substrate and depositing a conducting film on the resulting amorphous SiO_2 layer, forming the gate. However, the oxidation process always introduces defects at the Si-SiO_2 interface that critically affects the device characteristics. Right from the beginning of the integrated circuit era, there was a tendency to reduce the minimum feature length to achieve higher performances in terms of speed, component density, cost, etc. Increasing the wafer size and decreasing the device size, i.e., scaling, are reasonable approaches to produce faster devices consuming lesser power at a reduced cost. One of the most crucial elements that allows successful scaling of MOS devices is certainly the outstanding material and electrical properties of SiO_2. According to the International Technology Roadmap for Semiconductor (ITRS), the next generation Si-based MOS devices will require gate dielectric with thickness ~1 nm. But there are some fundamental scaling limitations with the ultrathin oxide. The tunneling current increases exponentially with reduction of gate oxide thickness that, in turn, results in unbearable power consumption and degrades the device performance. Furthermore, controlling the thickness uniformity of such an ultrathin film is another crucial issue. As a result, the scaling limit of SiO_2 depends on both the fundamental engineering physics and available technology. Presently, efforts are underway to replace SiO_2 by high-k gate dielectric materials. There are many contradictory requirements that need to be satisfied before a new material becomes acceptable to the industry for its use as the gate dielectric.

As per the ITRS report, the physical challenge is the most important challenge in the 21st century that could hold back CMOS from being further improved in the future. Identifying high-k dielectric materials with Effective Oxide Thickness (EOT) less than 0.5 nm and low leakage current are difficult challenges for MOSFET as well as advanced novel MOSFET structures like Tunneling FET, FinFET and Junctionless MOS Transistor, etc. With decreasing device dimensions, there is a dramatic increase of tunneling current, which has an adverse effect on the performance and functionality of CMOS applications. ITRS report shows the targets and requirements as a challenge for the CMOS technology maintaining the guiding scaling principle governed by the well-known Moore's law. The report also shows the high-k dielectric materials generation with the scaling process technologies timing, MPU/High-performance ASIC half pitch, gate length trends and timing, and industry. In fact, the Moore's law and the ITRS report strongly complement each other.

This book presents a broad review of SiO_2 material, including brief a historical note of Moore's law, followed by reliability issues of the SiO_2 based MOS transistor. Then it discusses the transition of gate dielectrics with an EOT ~ 1 nm and selection of high-k materials. A review of the different deposition techniques of different high-k films is also discussed. The challenges of high-k dielectric materials based MOS transistor is one of the most crucial issues in the evolving ITRS guidelines. High-k dielectrics theories (Quantum Tunneling effects and Interface Engineering theory) and applications of different novel MOSFET structures like Tunneling FET are also covered in this book.

The first chapter of the monograph presents a broad review of Moore's law. This includes the combined needs for digital and non-digital functionalities in an integrated system that is translated as a dual trend in the ITRS: miniaturization of digital functions ("More Moore") and functional diversification ("More than Moore"). It also includes a review of SiO_2 based MOSFET, including its limitations and further opportunities and challenges, together with a brief historical note of SiO_2 technology. Chapter 3 addresses the reporting transition to high-k dielectric materials with an effective oxide thickness below 1 nm, which also includes some structural properties of the high-k dielectric materials. The selection of the high-k dielectric materials for future generation of CMOS IC technology is presented in Chapter 4. Important issues of future CMOS technology in tunneling current and tunneling resistivity effect with the promising high-k gate dielectric material HfO_2 are described in Chapter 5. Analysis of interface charge densities with the high-k material Tantalum Pentoxide constitutes the next chapter. The issue of CMOS VLSI technology with the high-k gate dielectric materials is covered in Chapter 7. Advanced MOSFET structure: Tunnel FET, with its working, structure, and modeling is described in Chapter 8. The last chapter describes the Heusler compound material for optoelectronic, thermoelectric, and spintronic applications.

The challenges of high-k gate dielectric materials are one of the major concerns in the evolving ITRS. Based on the research experience, the authors feel that they address this topic with a timely, appropriate, interesting, motivating and resourceful monograph on both the fundamentals and successful integration of high-k dielectric materials in future IC technology.

CHAPTER 1

Moore's Law: In the 21st Century

N. P. MAITY* and RESHMI MAITY

Department of Electronics and Communication Engineering, Mizoram University (A Central University), Aizawl 796004, India

**Corresponding author. E-mail: maity_niladri@rediffmail.com*

ABSTRACT

This chapter describes the basic MOS structure and its development according to Moore's law and more. The combined requirement for digital and non-digital functionalities in an integrated system is interpreted as a double development in ITRS. The challenges for further development of advanced CMOS technologies are also explained

1.1 INTRODUCTION

"There is no reason anyone would want a computer in their home," this quote came from the lips of K. H. Olson, Chairman and Founder of Digital Equipment Corporation in 1977. At that period, the first generation microprocessor had only come to commercial market. Nowadays, getting at least one personal computer is natural in each household and small electronic gadgets such as smart phones, tablets etc., are very common and essential for every person. This terrific growth of the electronics industry is the consequences of several decades of enormous improvement in the manufacturing of large-scale integrated circuits (LSIs). The recent histrionic progresses of electronics and information technology owe to high-speed and smaller devices. This advancement has improved quality of life more freely and comfortably which are significantly contributed to the development of civilization and indispensable in modern societies.

Table 1.1 illustrates the 24 years (1990–2014) of global GDP growth versus worldwide semiconductor industry growth. It is very clear that the worldwide GDP growth strongly depends on the semiconductor industry growth (The McClean Report, 2015).

The progress of integrated circuits (ICs) has been accomplished with the downsizing (scaling) of metal oxide semiconductor devices. Scaling has facilitated IC manufactures to increase production exponentially, whereas decreasing cost at approximately the similar rate. In order to meet high-speed and high-density requirements of complementary metal-oxide-semiconductor (CMOS) ultra large-scale integrated circuits (ULSIs) downscaling of the transistor structures are being uninterruptedly attempted (Rao et al., 2014). The continuous downsizing in the feature dimensions has necessitated even thinner dielectric oxide layers (Yu et al., 2013; Maity et al., 2017, 2018, 2016, 2014, 2019). In advanced semiconductor technology, the transistor gate length will continue to shrink in order to achieve faster switching speed and higher device density. The corresponding impact will be the reduction of gate length and gate oxide thickness. However, with the advent of gate length technology of 45 nm and lower nodes, the requirement is to down scale the oxide thickness to an effective oxide thickness (EOT) of approximately 0.5 nm (Khaimar and Mahajan, 2013). As the size of the gate oxide thickness continues to scale down, the use of conventional SiO_2 is approaching physical and electrical limits such as problems of direct tunneling current and device reliability (Maity et al., 2016a, 2016b, 2017, 2014, 2018a, 2018b, 2019). So the present premises demand a metal-oxide-semiconductor (MOS) insulator with high dielectric constant, which can give a low leakage current and ultra thin oxide thickness (Dhar, 2013; Maity et al., 2019, 2018a, 2018b, 2017a, 2017b, 2016a, 2016b).

TABLE 1.1 Semiconductor Industry Growth Versus Worldwide GDP Growth (The McClean Report, 2015).

Year	Global GDP growth (percentage)	Semiconductor GDP growth (percentage)
1990	2.9	04.0
1991	1.4	08.0
1992	1.9	10.0
1993	1.6	29.0

TABLE 1.1 *(Continued)*

Year	Global GDP growth (percentage)	Semiconductor GDP growth (percentage)
1994	3.2	32.0
1995	2.9	42.0
1996	3.3	−9.0
1997	3.7	04.0
1998	2.5	−08.0
1999	3.4	19.0
2000	4.3	37.0
2001	1.7	−32.0
2002	2.1	01.0
2003	2.8	18.0
2004	4.2	28.0
2005	3.6	07.0
2006	4.1	09.0
2007	4.0	04.0
2008	1.4	−05.0
2009	−2.1	−10.0
2010	4.1	34.0
2011	2.9	02.0
2012	2.5	−03.0
2013	2.4	04.0
2014	2.9	08.0

1.2 THE MOS STRUCTURE

The MOS capacitor is the fundamental structure of the IC family. It is vital to study the MOS structure as it would benefit in understanding the IC fabrication and evaluating the electrical properties involving the MOS structures. It is the heart of the CMOS technology (Arora, 2007). Figure 1.1 represents the basic structure of a MOS transistor. Structurally, it is a simple capacitor—a parallel plate capacitor with semiconductor as one electrode and the metal as the other electrode. The insulator is generally an oxide layer of silicon. Primarily, gates were invariably made of aluminum

(Al) or polysilicon (poly-Si) and semiconductor unambiguously referred to the body. Such practice continues today, even though silicon dioxide (SiO_2) is not the only material used for the insulator. The silicon has an ohmic contact to provide an external electrical contact. The thickness of the insulator layer (oxide film) is denoted by t_{ox} and it determines the capacitance of the MOS system. V_{GB} is the voltage applied to the gate terminal of the MOS system with respect to the body to drive the MOS into its different operating regions according to the desired functionality. Body terminal is denoted by B.

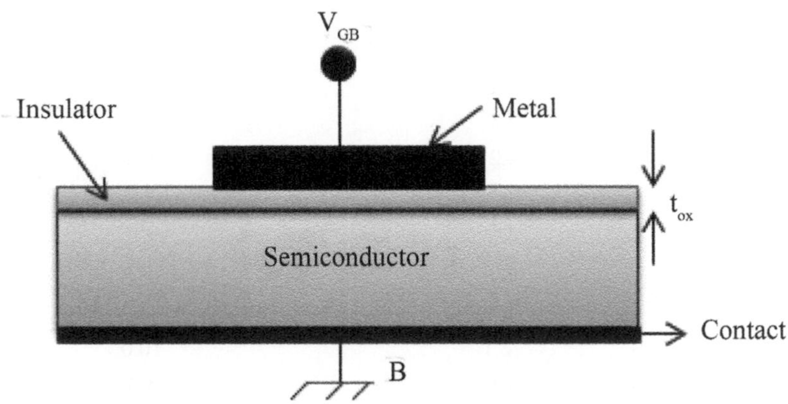

FIGURE 1.1 Basic MOS structure.

When there is no voltage applied in MOS structure ($V_{GB} = V_{FB}$), the bands are flat denoting flat-band condition. V_{FB} is called as flat-band voltage. There is no band bending under this condition. The metal semiconductor work function difference is:

$$\Phi_{ms} = \Phi_m - \Phi_s \tag{1.1}$$

For aluminum $\Phi_{ms} = -0.51V - \Phi_F$, for n^+ polysilicon gate $\Phi_{ms} = -0.56V - \Phi_F$, and for p^+ polysilicon gate $\Phi_{ms} = +0.56V - \Phi_F$ (Tsividis and McAndrew, 2011), where Φ_m is the metal work function, Φ_s is the semiconductor work function, and Φ_F is the Fermi potential.

The effect of the contact potential is not the only one that can result in a net concentration of charges in the body in the absence of applied potential. Parasitic charges also occur within the oxide layer as well as the oxide semiconductor interface (Gaddipati, 2004; Maity et al., 2014, 2016).

When, $V_{GB} \neq 0$, three cases may arise at the semiconductor surface and three kinds of potential drops are encountered in the MOS structure. Three cases are named as accumulation, depletion, and inversion according to the electron density on the semiconductor surface. Three kinds of potential drops are the potential drop across the oxide (ψ_{ox}), the surface potential $(\psi_{surface})$, and Φ_{ms}. So, we can write (Tsividis, 2011):

$$V_{GB} = \psi_{ox} + \psi_{surface} + \Phi_{ms} \tag{1.2}$$

Now, let us consider the charges in the MOS structure. Three kinds of charges present are the charge on the gate per unit area $(Q_G^{'})$, the interface charge per unit area $(Q_{it}^{'})$, and the charge in the semiconductor under the oxide $(Q_c^{'})$. These charges must balance one another for charge neutrality in the MOS structure (Tsividis, 2011):

$$Q_G^{'} + Q_{it}^{'} + Q_c^{'} = 0 \tag{1.3}$$

1.3 MOS SCALING AND MOORE'S LAW

To satisfy the demand for high-density ICs, the physical dimension of a MOS device must be on the nanoscale (Jhan et al., 2014). The fundamental governing principle of the tremendous development of MOS technology is the famous Moore's law which was first predicted in 1965 by Gordon Moore (Moore, 1965). This law stated that IC density and performance would double every 18 months. In 1975, he revisited his earlier prediction and delivered some critical visions into the technological drivers of the observed trends (Moore, 1975). It is significant to understand the important principles underlying Moore's law, since these permit us to improve insight into the future technology. These improvements would come from reduced transistor dimensions, increased transistor counts, and increased operating frequencies. The word Moore's law has come to denote to the continuous exponential development in the cost per function that can be accomplished on an IC. The importance of Moore's law lies less in the faithfulness of the rate of increment and more on its trend. The basic statement, of course, is that the increase in the fabrication cost of a chip is less than the increase in the number of components. The subsequent dramatic exponential reduction in cost per function is actually the driving force behind the progress in the semiconductor manufacturing industry and the information technology age. The primary influencing factor for the future growth of the CMOS technology is the cost per function (Moore, 1975). In 1995, G. Moore himself previously observed and mentioned that the IC manufacturing industry cannot carry on its fast exponential development forever. Ultimately, it is the technological limitation beyond which the Moore's law is no longer applicable (Isaac, 2000).

The evolution of simple transistor used in CMOS circuit was anticipated with outstanding accuracy in 1972 (Dennard et al., 1972) where a scientific scaling theory for higher and higher performance was projected. Today, one of the major limitations to the continuation of MOS transistor scaling theory is with tunneling through the gate oxide layer (Min et al., 2004; Kuo et al., 2004; Weber et al., 2006, Maity et al., 2016). At some point, leakage through gate oxide tunneling and others short channel effects limit the attainable channel length so that no further performance improvement can be achieved. Thus, gate oxide tunneling is a most important issue for limiting transistor scaling. Furthermore, study reveals that the leakage current can initiate damage, leading to previously unanticipated reliability issues in ultra thin gate dielectrics (Stathis and DiMaria, 1998). Reliability apprehensions might boundary gate oxide thickness to 1.5–2.0 nm (Isaac, 2000). Innovative device configurations or an alternative gate oxide material with a higher dielectric constant (high-k) will have to be established to accomplish a superior physical thickness and consequently reducing the tunneling current (leakage current) while retaining an effective lower oxide thickness.

"If Moore's law is simply a measure of the increase in the number of electronic devices per chip, then, Moore's law has much more time to go, probably decades," said Intel CTO Justin Rattner in recent times in an interview with Network World. Powel has shown the technology node from 130 nm to 22 nm versus the performance of the semiconductor chips (Powell, 2008). The gate terminal length possesses on shrinking as the technology node decreases. As what most researchers anticipated for previous few decades, the performance or the speed of the designed semiconductor chips must be increased as well.

However, it is observed from that the performance was, in contrast, decreased after technology node reached 65 nm. The primary and foremost reasons for this result are mainly gate tunneling current, power consumption, and heat sink (Wu et al., 2013). These factors are deciding the progress of the highly challenging consumer goods such as smartphone, laptop, and tablet. Powel (2008) shows the performance demands for processor in 2008. He has shown very clearly the relation between technology nodes with performance and length of gate. When technology nodes are decreasing the gate length also decreases, consequently performance also increases. Powel also very clearly mentioned the actual performance with projected and target performance. The global unit shipments of desktop PCs, notebook PCs versus smartphones and tablets for the year from 2005 to 2013 are predicted by Huberty and Gelblum in 2014 (Huberty and Gelblum, 2014). It is observed that the shipment of smartphones and tablets has been significantly increasing compared to desktop and notebook PCs after inflection point (smartphones + tablets > desktop PCs + notebook PCs). In 2011, it was nearly 500,000 mm, in 2012 it was 600,000 mm, and in 2013 it was above 700,000 mm. It sharply increases year by year. The complete worldwide mobile phone market is estimated to be worth $341.4 billion by 2015, whereas smartphones have occupied 75.8% of the overall demand in the same year (Wu et al, 2013).

1.4 "MORE MOORE" AND "MORE THAN MOORE"

The combined requirement for digital and nondigital functionalities in an integrated system is interpreted as a double development in the ITRS: miniaturization of digital functions ("More Moore") and functional diversification ("More than Moore"). The "More Moore" (Badaroglu et al., 2014; Mack, 2011; Schaller, 2002) domain is internationally defined as a challenge to further development of advanced CMOS technologies according to the Moore's law and reduce the associated cost per function. Nearly 70% of the total semiconductor components market (namely microprocessors, memories, and digital logics) is straight impacted by advanced CMOS miniaturization achieved in the More Moore domain. Challenges include, gate leakage, unsustainable power consumption, physical limits, manufacturing costs and process, and device variability. "More than Moore" (Jammy, 2010; Kazior, 2013; Bergeron, 2008) defines a set of technologies that enable nondigital microelectronic/nanoelectronic functions which is based on silicon technology do not necessarily scale with Moore's law. This trend is characterized by functional diversification of semiconductor electronics-based devices. "More than Moore" devices typically make available conversion of nondigital as well as nonelectronic information, such as mechanical, thermal, chemical, optical, acoustic, and biomedical functions, to digital data and vice versa. This combined essential for digital and nondigital functionalities in a product is depicted in Figure 1.2 (Arden, 2014).

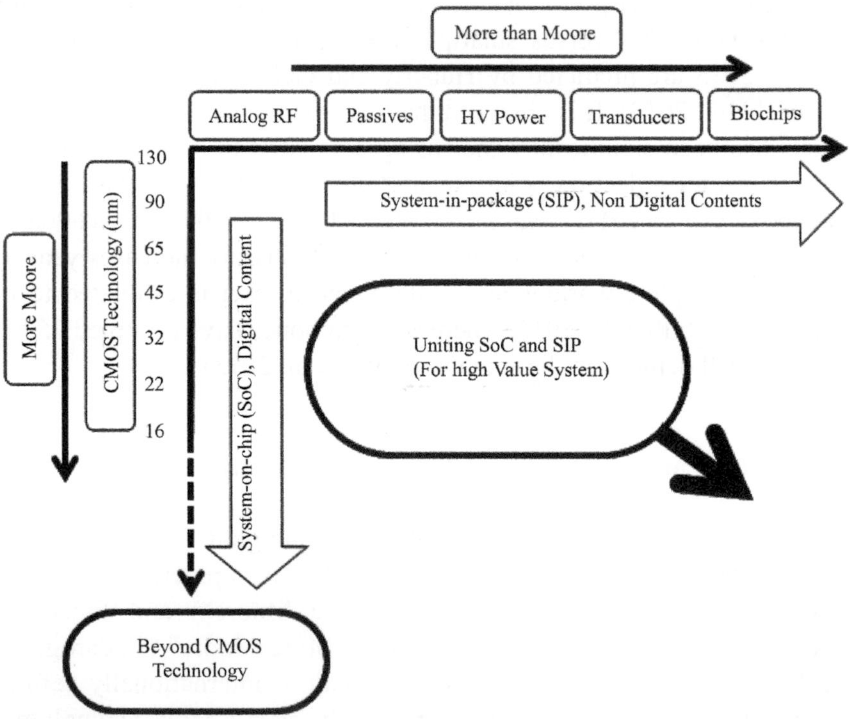

FIGURE 1.2 Moore's law and more (Adapted from Arden, 2014).

In detail, it is just like as the heterogeneous integration of digital and nondigital functionalities into a compacted system. The application field includes biomedical, communication, environment control, security,

MEMS/NEMS, and etc. Challenges includes bulky and incompatible with CMOS. So, to get more from Moore, "More than Moore" is a must (Jammy, 2010). Therefore "More Moore" might be observed as the brain of an intelligent condensed system, "More than Moore" denotes to its competences to interact with the outside world and the users.

KEYWORDS

- **Moore's law**
- **MOS**
- **MOS scaling**
- **MOSFET**
- **ITRS**
- **CMOS**

REFERENCES

Arden, W.; Brillouet, M.; Cogez, P.; Graef, M.; Huizing, B.; Mahnkopf, R. More than Moore: White Paper [Online] 2014. http//www.itrs.net.

Arora, N. *MOSFET Modeling for VLSI Simulation Theory and Practice,* 1st ed.; World Scientific: Singapore, 2007; pp 121–122.

Badaroglu, M.; Ng, K.; Salmani, M.; Kim, S.; Klimeck, G.; Chang, C.; Cheung, C.; Fukuzaki, Y. More Moore Landscape for System Readiness-ITRS2.0 Requirements. *Proc. IEEE Int. Conf. Comput. Design* **2014,** 147–152.

Bergeron, D. More than Moore. *Proc. IEEE CICC* **2008,** 25–26.

Dennard, R.; Kuhn, L.; Yu, H. Design of Micron MOS Switching Devices. *Proc. IEEE Int. Electron Device Meet.* **1972,** *18*, 168–170.

Dhar, J. C.; Mondal, A.; Singh, N. K.; Chinnamuthu, P. Low Leakage TiO_2 Nanowire Dielectric MOS Device using Ag Schottky Gate Contact. *IEEE Trans. Nanotechnol.* **2013,** *12*, 948–950.

Gaddipati, G. Characterization of HfO_2 Films for Flash Memory Applications. M.A. Thesis, Dept. Electrical Engineering, Univ. of South Florida, FL, USA, 2004.

Huberty, K.; Gelblum, E. The Morgan Stanley Research. Morgan Stanley: New York, USA, [Online] 2014. http//www.morganstanley.com.

Isaac, R. D. The Future of CMOS Technology. *IBM J. Res. Dev.* **2000,** *44*, 369–370.

Jammy, R. Life Beyond Si: More Moore and More than Moore. *Proc. IEEE Integr. Reliab. Workshop*, 2010.

Jammy, R. More Moore and More than Moore. *Proc. SEMATECH Symposium,* 2010.

Jhan, Y.; Wu, Y.; Lin, H.; Hung, M.; Chen, Y.; Yeh, M. High Performance of Fin-Shaped Tunnel Field-Effect Transistor SONOS Nonvolatile Memory with All Programming Mechanisms in Single Device. *IEEE Trans. Electron Devices* **2014,** *61*, 2364–2370.

Kazior, T. More than Moore: GaN HEMTs and Si CMOS Get it Together. *Proc. IEEE CSICS* **2013,** 1–4.

Khairnar, A.; Mahajan, A. M. Effect of Post-Deposition Annealing Temperature of RF-Sputtered HfO_2 Thin Film for Advanced CMOS Technology. *Solid State Sci.* **2013,** *15*, 24–28.

Kuo, C.; Hsu, J.; Huang, S.; Lee, L.; Tsai, M.; Hwu, J. High-k Al_2O_3 Gate Dielectrics Prepared by Oxidation of Aluminum Film in Nitric Acid Followed by High Temperature, Annealing. *IEEE Trans. Elect. Devices* **2004,** *51*, 854–8582.

Mack, C. A. Fifty Years of Moore's Law. *IEEE Trans. Semicond. Manuf.* **2011,** *24*, 202–207.

Maity, N. P.; Maity, R.; Thapa, R. K.; Baishya, S. Study of Interface Charge Densities Fro ZrO2 and HfO2 Based Metal Oxide Semiconductor Devices. *Adv. Mater. Sci. Eng.* **2014,** *2014*, 1–6.

Maity, N. P.; Maity, R.; Thapa, R. K.; Baishya, S. A. Tunneling Current Density model for Ultra Thin HfO2 High-k Dielectric Material Based MOS Devices. *Superlattices Microstruct.* **2016,** *95*, 24–32.

Maity, N. P.; Thakur, R. R.; Maity, R.; Thapa, R. K.; Baishya, S. Analysis of Interface Charge Densities for High- k Dielectric Materials Based Metal Oxide Semiconductor Devices. *Int. J. Nanosci.* **2016,** *15*, 1660011, 1–6.

Maity, N. P.; Maity, R.; Baishya, S. Voltage and Oxide Thickness Dependent Tunneling Current Density and Tunnel Resistivity Model: Application to High-k Material HfO2 Based MOS Devices. *Superlattices Microstruct.* **2017,** *111*, 628–641.

Maity, N. P.; Maity, R.; Thapa, R. K.; Baishya, S. Influence of Image Force Effect on Tunnelling Current Density for High-k Material ZrO2 Ultra Thin Films Based MOS Devices. J. Nanoelectron. *Optoelectron.* **2017,** *12* (1), 67–71.

Maity, N. P.; Maity, R.; Baishya, S. An Analytical Model for the Surface Potential and Threshold Voltage of a Double-Gate Hetero Junction Tunnel FinFET. *J. Comput. Electron.* DOI: org/10.1007/s10825-018-01279-5. Online Published 2018.

Maity, N. P.; Maity, R.; Baishya, S. A Tunneling Current Model with a Realistic Barrier for Ultra Thin High-k Dielectric ZrO2 Material based MOS Devices. *Silicon* **2018,** *10* (4), 1645–1652.

Maity, N. P.; Maity, R.; Maity, S.; Baishya, S. Comparative Analysis of the Quantum FinFET and Trigate FinFET Based on Modeling and Simulation. *J. Comput. Electron.* DOI: org/10.1007/s10825-018-01294-z). Online published 2019.

McClean, B. The McClean Report: A Complete Analysis of Forecast of the Integrated Circuit Industry. IC Insights Inc: Arizona, USA, [Online] 2015. http//www.icinsights.com/services/mccleanreport.

Min, B.; Devireddy, S.; Butler, Z.; Wang, F.; Zlotnicka, A.; Tseng, H.; Tobin, P. Low Frequency Noise in Sub Micrometer MOSFET with HfO_2, HfO_2/Al_2O_3 and $HfAlO_x$ Gate Stacks. *IEEE Trans. Electron Devices* **2004,** *51*, 1315–1322.

Moore, G. Cramming More Components onto Integrated Circuits. *Electronics* **1965,** *38*, 114–117.

Moore, G. Progress in Digital Integrated Electronics. *Proc. IEEE IEDM Tech. Dig.* **1975,** *21*, 11–13.

Powell, J. R. The Quantum Limit to Moore's Law. *Proc. IEEE* **2008,** *96*, 1247–1248.

Rao, A.; Dwivedi, Goswami, M.; Singh, B. R. Effect of Nitrogen Containing Plasma on Interface Properties of Sputtered ZrO_2 Thin Films on Silicon. *Mater. Sci. Semicond. Process.* **2014,** *19*, 145–149.

Schaller, R. Moore's Law: Past, Present and Future. *IEEE Spectr.* **2002**, *34*, 52–59.

Stathis, J.; DiMaria, D. Reliability Projection for Ultra Thin Oxides at Low Voltage. *Proc. Int. Electron Devices Meet.* **1998,** 167–170.

Tsividis, Y.; McAndrew, C. *Operation and Modeling of the MOS Transistor*, 3rd ed.; Oxford University Press: New York, 2011; pp 65–76.

Weber, O.; Damlencourt, J.; Andrieu, F.; Ducroquet, F.; Ernst, T.; Hartmann, J.; Papon, A.; Renoult, O.; Guillaumot, B.; Dilionibuas, S. Fabrication and Mobility Characteristics of SiGe Surface Channel PMOSFETs with a HfO_2/TiN Gate Stack. *IEEE Trans. Electron Devices* **2006**, *53*, 449–456.

Wu, J.; Shen, Y.; Reinhardt, K.; Szu, H.; Dong, B. A Nanotechnology Enhancement to Moore's law. *Appl. Comput. Soft Comput.* **2013**, *2013*, Article ID 426962, 13.

Yu, T.; Jin, C.; Zhang, H.; Zhuge, L.; Wu, Z.; Wu, X.; Feng, Z. Effect of Ta Incorporation on the Microstructure, Electrical and Optical Properties of $Hf_{1-x}Ta_xO$ High-k Film Prepared by Dual Ion Beam Sputtering Deposition. *Vacuum* **2013**, *92*, 58–64.

CHAPTER 2

SiO_2-Based MOS Devices: Leakage and Limitations

N. P. MAITY* and RESHMI MAITY

Department of Electronics and Communication Engineering, Mizoram University (A Central University), Aizawl 796004, India

Corresponding author. E-mail: maity_niladri@rediffmail.com

ABSTRACT

Leakage current becomes a great concern in deep submicron SiO_2 based CMOS technology consisting several major components. In this chapter every component has been described. The challenges and limitations of SiO_2 based MOS devices have been briefly explained. The prospects of high-k dielectric materials and their properties have been discussed.

2.1 INTRODUCTION

By flouting down abstractions and facing encounters in an integrated fashion from the devices up, designers could improve innovative schemes that at the end of the day go beyond von Neumann processing on silicon. For more than 50 years, thinkers have indicated the vision of a nanotechnology-based computer manufactured from the atoms or molecules up. For closely as long, many have anticipated the passing of the silicon microelectronic technology underlying conventional computing as we identify it. On the other hand long it takes silicon to respire its last gasp, the period is approaching when the computing industry essential appearance elsewhere for the means to sustain the prompt increase in capability that has permitted the improvement of an extensive range of novel applications, as well as great, cost-effective markets for memories and processors (Das, 2011).

2.2 MOS LEAKAGE COMPONENTS

Leakage current becomes a great concern in deep submicron CMOS technology consisting four major components. They are subthreshold leakage current, reverse bias junction leakage, leakage current formed by drain-induced barrier lowering (DIBL), and gate tunneling leakage. There are still additional leakage components, like gate-induced drain leakage (GIDL), subthreshold leakage, and bulk punch through current. Figure 2.6 shows that the overall leakage current can be divided into several components for a MOS transistor (Bergeron et al., 2008).

2.2.1 Subthreshold Leakage Current (I_{sub})

Applied voltage has been scaled down to possess dynamic power consumption under control. To keep the higher drive current proficiency, the threshold voltage (V_T) has to be scaled as well. On the other hand, the V_T scaling results in increasing I_{sub}. Temperature is also a most significant parameter for I_{sub}. I_{sub} decreases in an exponential manner with increasing threshold voltage and decreasing temperature. Fundamentally, shorter effective channel length, longer transistor width, lower channel doping will reduce V_T and increase I_{sub}.

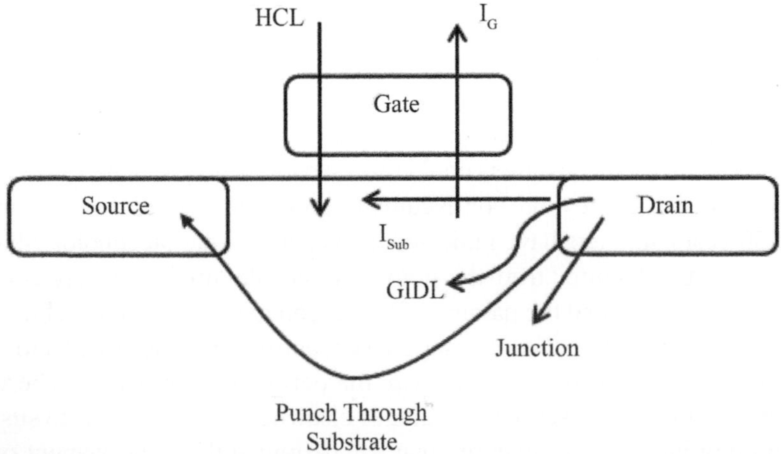

FIGURE 2.1 Schematic descriptions of the different leakage currents and mechanisms in deep submicron MOS transistor (Bergeron et al., 2008).

In addition, the minority carrier concentration is nearly zero in weak inversion situation, and the channel has definitely not horizontal electric field, but a slight longitudinal electric field perform sowing to the drain to source voltage. In this condition, the carriers travel by diffusion between the source and drain terminal of MOS transistor. Accordingly, I_{sub} is also controlled by diffusion current. Considering the BSIM MOS transistor model the I_{sub} for a MOS transistor can be given by:

$$I_{sub} = I_0 \exp\left[{(V_{GS} - V_T)}\big/{nV_{th}} \right]\left[1 - \exp\left(-{V_{DS}}\big/{V_{th}} \right) \right] \quad (2.1)$$

Where, $I_0 = \left[\left\{ W \mu C_{ox} V_{th}^{\,2} \exp(1.8) \right\} / L \right]$, V_{DS} and V_{GS} are denoted as the drain to source and gate to source voltage respectively. W and L are represented as the effective transistor width and length, respectively. C_{ox} is the oxide capacitance, μ is the carrier mobility, n is the subthreshold swing coefficient, and V_{th} is represented as the thermal voltage at temperature T (Shauly, 2012; Sheu et al., 1987; Anjana and Somkuwar, 2013; Yadav and Akashe, 2012; Kao et al., 2002; Deepaksubramanyan and Nunez, 2007; Powell et al., 2001; Narendra et al., 2004).

2.2.2 Junction Leakage Current

Fundamentally, junction leakage current occurs due to the reverse bias PN junction which comprises of two mechanisms. First one is diffusion of minority carriers nearby the edge of the depletion region and other one is the generation of electron hole pair (EHP) in the depletion region of reverse bias junction. Junction leakage does not contribute more in the total leakage current. The junction leakage current is 7×10^{-12} A for minimal operating mode when substrate doping is 5×10^{15} cm $^{-3}$ (p-type), drain doping is 9×10^{19} cm $^{-3}$ (n-type), drain PN junction area is 5 μm × 1.5 μm and drain voltage is 1.8 V (Kudo et al., 2014; Wang., 2014; Tarun et al., 2003; Uejima and Hase, 2011; Eneman et al., 2007; Maji et al., 2013; Semenov et al., 2003).

2.2.3 DIBL Effect

D in DIBL term refers to V_{DS} (drain voltage) and B in DIBL term refers to V_T (i.e. the barrier). Therefore, "Drain-Induced Barrier Lowering" means "V_{DS} induced V_T lowering". For submicron MOS transistors, the threshold voltage is acknowledged to decrease linearly with the increase of drain terminal voltage. The foremost mechanism that decreases the V_T is the decrease of the potential barrier in the depletion region below the gate terminal owing to the applied drain voltage. This is called DIBL effect. The V_T model including DIBL effect is given by:

$$V_T = V_T^0 - \zeta_{DIBL} V_{DS} \tag{2.2}$$

Where, V_T^0 is the threshold voltage at zero drain voltage and ζ_{DIBL} is the DIBL coefficient and it is given by the change in threshold voltage (ΔV_T) and divided by change in drain voltage (ΔV_{DS}) with a negative sign, that is:

$$\zeta_{DIBL} = -\left[{}^{\Delta V_T}\!/\!{}_{\Delta V_{DS}} \right] \tag{2.3}$$

It has been evaluated the DIBL coefficient is equal to 15 mV/V for n-MOS transistor with t_{ox} = 4 nm and channel doping 4 × 10^{17} cm $^{-3}$ (Semenov et al., 2003; Liu et al., 2011; Ghitani, 1999; Ng et al., 1993; Bukkawar and Sarwade, 2012; Barua et al., 2014; Bordollo et al., 2014).

2.2.4 Gate Tunneling Leakage

The aggressive MOS device scaling in nanometer regime increases the short-channel effects (SCEs). To regulate the SCEs, oxide thickness (insulator thickness) is essential to become thinner and thinner to every technology generation. Consistent scaling of oxide thickness, in turn, gives rise to higher electric field, resulting in a high direct tunneling current through transistor gate insulator (Maity et al., 2016, 2017). For thick oxide layer (t_{ox} ~ 3–10 nm) the current is controlled by Fowler–Nordheim tunneling, while for the ultra thin oxide layer (t_{ox}< 3 nm) at voltage below about 3 V the current is due to direct quantum mechanical tunneling.

The gate tunneling current increases as the physical thickness of the gate oxide is decreased. In quest of increased control over the channel region, the oxide thickness has been continuously scaled down which has become now as small as few atomic layers only. Generally, there are two types of tunneling phenomenon (1) Fowler–Nordheim (F–N) tunneling and (2) Direct tunneling. The tunneling probability depends on many factors including width and height of the barrier and its structure. In the event of F–N tunneling, electrons tunnel through a triangular potential barrier where as in the situation of direct tunneling, electrons tunnel through a trapezoidal barrier. The gate leakage current has been modeled by Roy et al. (Roy et al., 2003) and it is given by:

$$I_{gate} = W \times L \times u\left(\frac{\psi_{ox}}{t_{ox}}\right)^2 \exp\left[\frac{-v\left[1-\left(1-\frac{\psi_{ox}}{\phi_{OX}}\right)\right]}{\frac{\psi_{ox}}{t_{ox}}}\right] \tag{2.4}$$

Where, $u = \left(q^3/16\pi^2 h\phi_{OX}\right)$, $v = \left[\left(4\pi\sqrt{2m^*}\psi_{ox}^{3/2}\right)/3hq\right]$, m^* is the effective mass of the tunneling particle, ϕ_{OX} is the tunneling barrier height, q is the electronic charge, and h is the $1/2\pi$ times Plank's constant (Roy et al., 2003; Choi and Dutton, 2004; Momose, 1998).

2.2.5 *GIDL Current*

GIDL is initiated by high field effect in the drain terminal junction of MOS transistor. Thinner oxide layer and higher supply voltage increase GIDL current. Several mechanisms has described the characteristics of GIDL current like the direct band to band tunneling model and the band trap band tunneling model. On the other hand, the band to band tunneling was recognized as the foremost leakage mechanism based on qualitative agreement between experiment results and analytical model results. A simple 1-D band to band tunneling current model was presented by Chen et al. which predicts the GIDL current as:

$$I_{GIDL} = A \times E_s \exp\left[-\frac{B}{E_s}\right] \tag{2.5}$$

Where, A is a constant and defined as:

$$A = \left[\frac{q^2\sqrt{m_r}}{18\pi h^2 E_{gap}^{3/2}}\right] \tag{2.6}$$

B is also a constant and defined as:

$$B = \left[\frac{\pi\sqrt{m_r}\,E_{gap}^{3/2}}{2\sqrt{2}q\hbar}\right] = 21.3MV/cm \tag{2.7}$$

with $m_r = 0.2\, m_0$. m_0 is the electron effective mass, E_{gap} is the direct energy gap of silicon (~ 3.5 eV), $\hbar$ is Plank's constant, and E_s is the vertical electric field at semiconductor surface (Semenov, 2003; Chen et al., 1987; Choi et al., 2003).

2.2.6 *Punch through Leakage*

It occurs when the depletion regions from source and drain merge with each other and a space charge limited current flows between source and drain. This current cannot be controlled with the help of gate voltage whereas it may be controlled by changing the drain biasing voltage. Channel is highly doped to prevent the punch through effect but due to this high doping there is an increase in tunneling current between source/drain and channel junctions. In the off mode (lesser V_{DS}), the minimum potential of the electric field is situated nearby to the middle of the gate terminal, for the reason that the source and drain potentials have comparable magnitude. Increasing V_{DS} modifies the minimum potential toward the source terminal and reduces the source potential barrier, so that the drain to source current is increased. This current is the bulk punch through current. For large V_{DS} the current becomes space charge limited (SCL). The elementary model for SCL current has given by Vanstraelen et al., which gives:

$$I_{SCL} \approx \left[\frac{9\varepsilon_s \mu d}{8L^3}\right]\left[V_{DS} - V_{DS}^{SCL}\right]^2 \quad (2.8)$$

Where, V_{DS}^{SCL} is defined as:

$$V_{DS}^{SCL} = \left(\frac{2}{k_d}\right)\left(\frac{KT}{q}\right)\ln\left(\frac{N_d}{n_i}\right) \quad (2.9)$$

and $k_d = (\varepsilon_s)/(L\chi C_{ox})$, χ is a geometry-dependent fitting parameter, d is the channel cross section, and N_d is source/drain doping (Semenov, 2003; Koyanagi et al., 1987; Lee et al., 2001).

2.2.7 Hot Carrier Injection (HCI)

HCI is a phenomenon in semiconductor electronic devices in a short-channel MOS transistor. It is due to high electric field neighboring the semiconductor-oxide interface electrons and hole can increase sufficient energy from the electric field to cross the interface potential barrier and enter into the oxide region. This consequence is known as HCI. Hot electrons are electrons that have gained very high kinetic energy after being accelerated by a high electric field in areas of high intensities within a MOS device. The injection from semiconductor to oxide is more probable for electrons than holes as electrons have a lower effective mass than that of holes. HCI is one of the mechanisms that adversely affects the reliability of the semiconductor of solid state electronics devices (Taur and Ning, 1998; Ghobadi et al., 2009; Fang et al., 2014).

2.3 SHORT CHANNEL EFFECTS

As the channel length is reduced to increase the operation speed and the number of transistors per chip, the so-called SCE arises. The SCEs are contributed to two physical phenomena: (1) the constraint imposed on electron drift characteristics in the channel region and (2) the alteration of the threshold voltage owing to the shortening channel length. Scaling has led to the devices with smaller gate lengths as shown in Figure 2.2 of which behavior cannot be described by the first order device equations.

The smaller gate sizes are driven by the continuous improvement in the lithographic processes. This result into higher drive currents and hence faster circuits may be expected if the parasitic capacitances are also assumed to be scaled simultaneously (Rechem et al., 2009; Suzuki, et al., 1993; Boucart and Ionescu, 2007; Iwai, 2009; Yu et al., 2008).

In a short-channel device, the channel length is of the same order to the depletion layers of the source and drain junctions (Kuo et al., 2002; Yuan and Yu, 2009; Agarwal et al., 1996; Chen and Kuo, 1996). The second order effects like quantum mechanical effects, threshold voltage variations with channel length, DIBL due to the charge sharing and doping effects, extrinsic resistances, etc. must be taken into account. The combined effect of these second order effects called SCEs make it difficult to have a clear cut idea about the device design parameters.

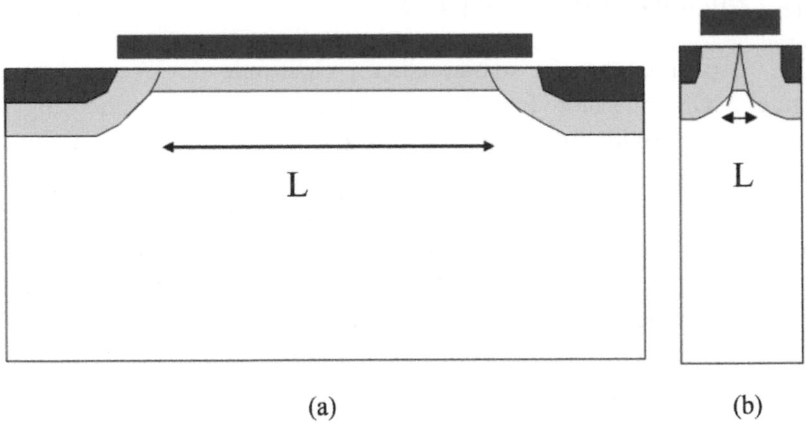

FIGURE 2.2 (a) Long-channel MOSFET and (b) short-channel MOSFET.

2.4 MOS SCALING AND LIMITATION OF SiO_2

By ITRS report, the physical challenge is the most important challenge in 21st century that could hold back CMOS from further improvement in future. Identifying high-k dielectrics materials with effective oxide thickness (EOT) < 0.5 nm and low leakage current are then difficult challenges (ITRS, 2013). These are due to dramatic increment of tunneling currents as the devices are becoming smaller and smaller (Vishnoi and Kumar, 2014; Cho, 2014) thus impacting the performance and functionality of

CMOS devices (Chen et al., 2014). The ITRS report (ITRS, 2012), shows the targets and requirements as a challenge for the CMOS technology maintaining the guiding scaling principle governed by the well-known Moore's law. This report also shows the generation of high-k dielectric materials, channel materials, and electrostatic control structure with the scaling process technologies (see Fig. 2.3 and Table 2.1) (He and Sun, 2012). In fact, the Moore's law and the ITRS report strongly complement each other.

With the ever increasing demands for speed and density of the silicon-integrated circuits, MOS device scaling has become a primary concern of the semiconductor manufacturing industry. The exceptional evolution of silicon material is promptly approaching a saturation point where device fabrication can no longer be simply scaled to increasingly smaller sizes. The origin of this saturation is indicated, where not only the gate leakage current density but also the corresponding gate dielectric thicknesses are plotted as a function of time. Thinning of the gate dielectric compulsory by scaling guidelines and essential for reaching the next generations of integrated devices will be the source of unacceptable high leakage current arising from electron tunneling through the SiO_2.

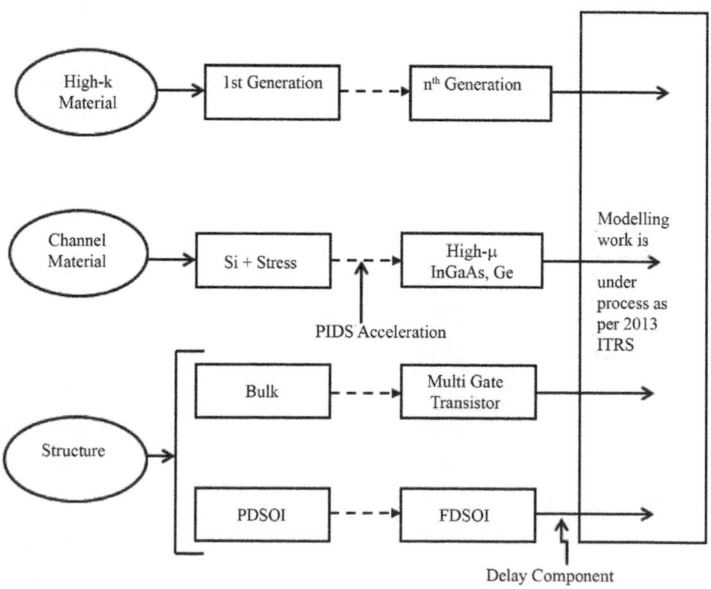

FIGURE 2.3 High-k material, channel material, and structure trends (He and Sun, 2012).

The scaling limit of SiO_2 depends on both the elementary physics and technology. Even though the scaling of MOS devices utilizing SiO_2 as gate material has reached its limit, there is a prospect in thinning the electrical oxide thickness further by using high-k dielectric materials and/or metal gates to accomplish a greater physical thickness and therefore reducing the direct tunneling current while retaining an effective low oxide thickness (Bohr, 2002; Wu et al., 2013; Yeo et al., 2001; Thompson et al., 1998, Maity et al., 2016, 2017, 2018).

TABLE 2.1 Scaling Process Technologies Timing, MPU/High-Performance ASIC Half Pitch and Gate Length Trends and Timing (He and Sun, 2012).

ITRS	Year	Technology node
2011 ITRS flash memory	2007	54 nm
	2009	45 nm
	2012	22 nm
	2015	15 nm
	2018	11 nm
	2021	11 nm
	2022–24	8 nm
2011 ITRS DRAM	2007	68 nm
	2009	45 nm
	2012	32 nm
	2015	22 nm
	2018	16 nm
	2021	11 nm
	2022–24	8 nm
MPU/hpASIC	2007	45 nm
	2009	32 nm
	2012	22 nm/20 nm
	2015	16 nm/14 nm
	2018	11 nm/10 nm
	2021	8 nm/7 nm
2011 ITRSMPU hpASIC	2007	76 nm
	2009	54 nm
	2012	38 nm
	2015	27 nm

TABLE 2.1 *(Continued)*

ITRS	Year	Technology node
	2018	19 nm
	2021	13 nm
2011 ITRS GLPr	2007	54 nm
	2009	47 nm
	2012	35 nm
	2015	28 nm
	2018	20 nm
	2021	14 nm
2011 ITRS GLPh	2007	32 nm
	2009	29 nm
	2012	24 nm
	2015	20 nm
	2018	15 nm
	2021	12 nm

To date, however, there is no particular high-k material that is competent to satisfy all the requirements for an ideal gate oxide for MOS devices. There arc many contradictory requirements need to be satisfied before a new material is acceptable to the semiconductor industry for its use as gate dielectric. Several insulating materials in the lanthanum group exhibiting higher dielectric constant, as shown in Table 2.2 (Persson, 2004; Silvaco, 2013; Maity et al., 2016, 2017), in the range of 7–30, have been under investigation as prospective replacement for SiO_2 and show better results.

The choice of a material with higher dielectric constant than SiO_2 can give the same equivalent oxide thickness with a higher physical thickness. In an attempt to replace conventional SiO_2 with new high-k materials, Al_2O_3, HfO_2, and ZrO_2 have received considerable attention as quite promising materials (Gaddipati, 2004; Jeong et al., 2005; Chen et al., 2002; Wilk et al, 2001; Fan et al., 2002; Cass'e et al., 2006; Sohn et al., 2011; Wu et al., 2012; Cho et al., 2007a; Sahoo and Oates, 2014, Maity et al., 2017b, 2019, 2014, 2018).

TABLE 2.2 Properties of High-k Dielectric Materials (Persson, 2004; Maity et al., 2016, 2017; Silvaco, 2013).

Material	k	Bandgap (eV)	Crystal structure
SiO_2	3.9	9	Amorphous
Si_3N_4	7	5.3	Amorphous
Al_2O_3	9.3	8.8	Amorphous
$HfSiO_4$	15	6	Amorphous
Y_2O_3	15	6	Cubic
$ZrSiO_4$	15	6	Amorphous
HfO_2	22	6	Monoclinic, tetragonal
ZrO_2	22	5.8	Monoclinic, tetragonal
Ta_2O_5	26	4.4	Orthorhombic
La_2O_3	30	6	Hexagonal, cubic

However, high-k materials often suffer from poor oxide dielectric-semiconductor interface quality and are repeatedly associated with lower dielectric breakdown voltages along with decreased lifetimes. In addition, with the introduction of high-k materials there occurs formation of interface charges. Owing to scaling, it has grown into more significant to consider the effect of generated traps in Si–SiO_2 junction. The use of high-k materials produces a large number of interface traps at the surface and oxide trap charges in the gate dielectric bulk of MOS transistors, which results in deprivation of device electrical characteristics (Sahoo and Oates, 2014). As the oxide thickness is reduced these interface trap charges gradually become important (Sahoo and Oates, 2014). This is because, even though high-k materials compromise higher capacitance, they repeatedly suffer from poor electrical quality of oxide-semiconductor interface and is often associated with lower dielectric breakdown voltages and decreased lifetimes. Moreover, scaling leads to generation of significant trap charges in oxide-semiconductor junction. During the last three decades, the gate oxide thickness was so large that this phenomenon was not noticeable; however, nowadays the effects of these interface charge states can no longer be ignored (Zhu, et al., 2002; Kerber et al., 2003; Hoogeland, et al., 2009).

KEYWORDS

- **nanoelectronics**
- **MOS**
- **ITRS**
- **short-channel effect**
- **leakage**
- **tunneling current**
- **CMOS**
- **high-k**

REFERENCES

Agrawal, B.; De, V. K.; Meindl, J. D. Device Parameter Optimization for Reduced Short Channel Effects in Retrograde Doping MOSFET's. *IEEE Trans. Electron Devices* **1996,** *43*, 365–368.

Anjana, R.; Somkuwar, A. Analysis of Sub-Threshold Leakage Reduction Techniques in Deep Submicron Regime for CMOS VLSI Circuits. *Proc. IEEE ICEVENT* **2013**, 1–3.

Barua, P.; Jafar, I.; Sengupta, P.; Noor, M. Mixed FBB and RBB Low Leakage Technique for High Durable CMOS Circuit. *Proc. IEEE ICIEV* **2014,** 1–5.

Bergeron, D. More than Moore. *Proc. IEEE CICC,* **2008,** 25–26.

Bohr, M. T. Nanotechnology Goals and Challenges for Electronic Applications. *IEEE Trans. Nanotechnol.* **2002,** *1*, 56–62.

Bordollo, C.; Teixeira, F.; Silveira, M.; Martino, J.; Agopian, P.; Simoen, E.; Claeys, C. The Effect of X-Ray Radiation on DIBL for Standard and Strained Triple Gate SOI MuGFETs. *Proc. IEEE ICCDCS* **2014,** 1–4.

Boucart, K.; Mihai Ionescu, A. Length Scaling of the Double Gate Tunnel FET with a High-K Gate Dielectric. *Solid-State Electron.* **2007,** *51*, 1500–1507.

Bukkawar, S.; Sarwade, N. Low Leakage Nano Scaled Body on Insulator FinFET with Underlap and High-k Dielectric. *Proc. IEEE ICCICT* **2012,** 1–5.

Cass´e, M.; Thevenod, L.; Guillaumot, B.; Tosti, L.; Martin, F.; Mitard, J.; Weber, O.; Andrieu, F.; Ernst, T.; Reimbold, G.; Billon, T.; Mouis, M.; Boulanger, F. Carrier Transport in HfO_2/Metal Gate MOSFETs: Physical Insight Into Critical parameters. *IEEE Trans. Electron Devices* **2006,** *53*, 759–768.

Chen, J.; Chan, T.; Chen, I.; Ko, P.; Hu, C. Sub-Breakdown Drain Leakage Current in MOSFET. *IEEE Electron Device Lett.* **1987,** *8*, 515–517.

Chen, S.; Kuo, J. B. Deep Sub Micrometer Double-Gate Fully-Depleted SO1 PMOS Devices: A Concise Short-Channel Effect Threshold Voltage Model Using a Quasi-2-D Approach. *IEEE Trans. Electron Devices* **1996,** *43*, 1387–1393.

Chen, S.; Lai, C.; Chan, K.; Chin, A.; Hsieh, J.; Liu, J. Frequency-Dependent Capacitance Reduction in High-k $AlTiO_X$ and Al_2O_3 Gate Dielectrics from IF to RF Frequency Range. *IEEE Electron Device Lett.* **2002,** *23*, 203–205.

Chen, Y.; Fan, M.; Hu, V. P.; Su, P.; Chuang, C. T. Evaluation of Sub-0.2 V High-Speed Low-Power Circuits Using Hetero-Channel MOSFET and Tunneling FET Devices. *IEEE Trans. Circuits Syst. I: Regul. Pap.* **2014,** *61*, 3339–3347.

Cho, H. J.; Kim, Y. D.; Park, D. S.; Lee, E.; Park, C. H.; Jang, J. S.; Lee, K. B.; Kim, H. W.; Ki, Y. J.; Han, I. K.; Song, Y. W. New TIT Capacitor with $ZrO_2/Al_2O_3/ZrO_2$ Dielectrics for 60 nm and Below DRAMs. *Solid State Electron.* **2007,** *51*, 1529–1533.

Cho, W.; Gupta, S. K.; Roy, K. Device Circuit Analysis of Double Gate MOSFETs and Schottky Barrier FETs: A Comparison Study for Sub-10 nm Technologies. *IEEE Trans. Electron Devices* **2014,** *61*, 4025–4031.

Choi, Y.; Ha, D.; King, T.; Bokor, J. Investigation of Gate-Induced Drain Leakage (GIDL) Current in Thin Body Devices: Single-Gate Ultra thin Body, Symmetrical Double-Gate, and Asymmetrical Double-Gate MOSFETs. *Jpn. J. Appl. Phys.* **2003,** *42*, 2073–2076.

Choi, C.; Dutton, R. W. Gate Tunneling Current and Quantum Effects in Deep Scaled MOSFETs. *J. Semicond. Technol. Sci.* **2004,** *4*, 27–31.

Das, S. Transitioning from Microelectronics to Nanoelectronics. *IEEE Computer Society* **2011,** *44*, 18–19.

Deepak Subramanyan, B.; Nunez, A. Analysis of Sub-Threshold Leakage Reduction in CMOS Digital Circuit. *Proc. IEEE MWSCAS* **2007,** 1400–1404.

Eneman, G.; Casain, O.; Simoen, E.; Brunco, D.; Jaeger, B.; Satta, A.; Nicholas, G.; Claeys, C.; Meuris, M.; Heyns, M. Analysis of Junction Leakage in Advanced Germanium P+/n Junctions. *Proc. IEEE ESSDERC* **2007,** 454–457.

Fan, Y.; Nieh, R. E.; Lee, J. C.; Lucovsky, G.; Brown, G.; Register, L.; Banerjee, S. K. Voltage- and Temperature-Dependent Gate Capacitance and Current Model: Application to ZrO_2 n-channel MOS capacitor. *IEEE Trans. Electron Devices* **2002,** *49*, 1969–1978.

Fang, J.; Sapatnekar, S. Incorporating Hot-Carrier Injection Effects into Timing Analysis for Large Circuits. *IEEE Trans. VLSI Syst.* **2014,** *22*, 2738–2751.

Gaddipati, G. *Characterization of HfO_2 Films for Flash Memory Applications*. M. A. Thesis, Dept. Electrical Engineering, Univ. of South Florida, FL, USA, 2004.

Ghitani, H. E. DIBL Coefficient in Short Channel NMOS Transistors. *Proc. IEEE NRCS* **1999,** D4(1)–D4(5).

Ghobadi, N.; Afzali-Kusha, A.; Asl-Soleimani, E. Analytical Modeling of Hot Carrier Injection Induced Degradation in Triple Gate bulk FinFETs. *Proc. IEEE ASQED* **2009,** 28–34.

He, G.; Sun, Z. *High-k Gate Dielectrics for CMOS Technology*, 1st ed.; Wiley/VCH: Welnheim, Germany, 2012.

Hoogeland, D.; Jinesh, K.; Roozeboom, F.; Besling, W.; Sanden, M.; Kessels, W. Plasma Assisted Atomic Layer Deposition of TiN /Al_2O_3 Stacks Formetal Oxide-Semiconductor Capacitor Applications. *J. Appl. Phys.* **2009**, *106*, Article ID-114107.

International Technology Roadmap for Semiconductors (2012). [Online] 2012. http://www.itrs.net/reports.html.

Iwai, H. Roadmap for 22 nm and Beyond. *Microelectron. Eng.* **2009,** *86*, 1520–1528.

Jeong, S.; Bae, I.; Shin, Y.; Lee, S.; Kwak, H.; Boo, J. Physical and Electrical Properties of ZrO_2 and YSZ High-k Gate Dielectric Thin Films Grown by RF Magnetron Sputtering. *Thin Solid Films* **2005,** *475*, 354–358.

Kao, J.; Narendra, S.; Chandrakasan, A. Sub-Threshold Leakage Modeling and Reduction Techniques [IC CAD tools]. *Proc. IEEE ICCAD* **2002,** 141–148.

Kerber, A.; Cartier, E.; Pantisano, L.; Degraeve, R.; Kauerauf, T.; Kim, Y.; Hou, A.; Groeseneken, G.; Maes, H.; Schwalke, U. Origin of the Threshold Voltage Instability in SiO_2/HfO_2 Dual Layer Gate Dielectrics. *IEEE Electron Device Lett.* **2003,** *24*, 87–89.

Koyanagi, M.; Lewis, A.; Martin, R.; Huang, T.; Chen, J. Hot-Electron-Induced Punch through (HEIP) Effect in Submicrometer PMOSFET's. *IEEE Trans. Electron Devices* **1987,** *34*, 839–844.

Kudo, S.; Hirose, Y.; Yamaguchi, T.; Kashihara, K. Analysis of Junction Leakage Current Failure of Nickel Silicide Abnormal Growth Using Advanced Transmission Electron Microscopy. *IEEE Trans. Semicond. Manuf.* **2014,** *27*, 16–21.

Kuo, J. B.; Yuan, K.; Li, S. Compact Threshold-Voltage Model for Short-Channel Partially-Depleted (PD) SOI Dynamic-Threshold MOS (DTMOS) Device. *IEEE Trans. Electron Devices* **2002,** *49*, 190–196.

Lee, J.; Lee, S.; Jeong, G.; Chung, T.; Kim, K. Investigation of pMOSFET Hot Electron Induced Punch Through (HEIP) is Shallow Trench Isolation. *Proc. IEEE 31st Eur. Solid-State Device Res. Conf.* **2001,** 467–470.

Liu, Z.; Hu, Z.; Zhang, Z.; Shao, H.; Ning, B.; Chen, M.; Bi, D.; Shichang, Z. Total Ionizing Dose Enhanced DIBL Effect for Deep Submicron NMOSFET. *IEEE Trans. Nucl. Sci.* **2011,** *58*, 1324–1331.

Maity, N. P.; Maity, R.; Thapa, R. K.; Baishya, S. Study of Interface Charge Densities for ZrO_2 and HfO_2 Based Metal Oxide Semiconductor Devices. *Adv. Mater. Sci. Eng.* **2014,** *2014*, 1–6.

Maity, N. P.; Maity, R.; Thapa, R. K.; Baishya, S.A Tunneling Current Density model for Ultra Thin HfO_2 High-k Dielectric Material Based MOS Devices. *Superlattices Microstruct.* **2016,** *95*, 24–32.

Maity, N. P.; Maity, R.; Baishya, S. Voltage and Oxide Thickness Dependent Tunneling Current Density and Tunnel Resistivity Model: Application to High-k Material HfO2 Based MOS Devices. *Superlattices Microstruct.* **2017,** *111*, 628–641.

Maity, N. P.; Maity, R.; Thapa, R. K.; Baishya, S. Influence of Image Force Effect on Tunnelling Current Density for High-k Material ZrO2 Ultra Thin Films Based MOS Devices. *J. Nanoelectron. Optoelectron.* **2017,** *12* (1), 67–71.

Maity, N. P.; Maity, R.; Baishya, S. A. Tunneling Current Model with a Realistic Barrier for Ultra Thin High-k Dielectric ZrO2 Material based MOS Devices. *Silicon* **2018,** *10* (4), 1645–1652.

Maity, N. P.; Maity, R.; Maity, S.; Baishya, S.; Comparative Analysis of the Quantum FinFET and Trigate FinFET Based on Modeling and Simulation. *J. Comput. Electron.* [Online] 2019. DOI: org/10.1007/s10825-018-01294-z.

Maji, D.; Liao, P.; Lee, Y.; Shih, J.; Chen, S.; Gao, S.; Lee, J.; Wu, K. A Junction Leakage Mechanism and its Effects on Advanced SRAM Failure. *Proc. IEEE IRPS* **2013,** 3E1.1–3E1.5.

Momose, H. Study of the Manufacturing Feasibility of 1.5-nm Direct-Tunneling Gate Oxide MOSFET's: Uniformity, Reliability, and Dopant Penetration of the Gate Oxide. *IEEE Trans. Electron Devices* **1998**, *45*, 691–700.

Narendra, S.; De, V.; Borker, S.; Antoniadis, D.; Chandrakasan, A. Full Chip Sub-Threshold Leakage Power Prediction and Reduction Technique for Sub 0.18 μm CMOS. *IEEE J. Solid State Circuits* **2004**, *39*, 501–510.

Ng, K.; Eshraghi, S.; Stanik, T. An Improved Generalized Guide for MOSFET Scaling. *IEEE Trans. Electron Devices* **1993,** *40*, 1895–1897.

Persson, S. Modeling and Characterization of Novel MOS Devices. Ph.D. Thesis, Solid State Devices Laboratory, Royal Institute of Technology, Stockholm, Sweden, 2004.

Powell, M.; Yang, S.; Falsafi, B.; Roy, K.; Vijaykumar, T. Reducing Leakage in a High Performance Deep Submicron Instruction Cache. *IEEE Trans. VLSI Syst.* **2001,** *9*, 77–89.

Rechem, D.; Latreche, S.; Gontrand, C. Channel Length Scaling and the Impact of Metal Gate Work Function on the Performance of Double Gate-Metal Oxide Semiconductor Feld-Effect Transistors. *Pramana—J. Phys.* **2009,** *72*, 587–599.

Roy, K.; Mukhopadhyay, S.; Mahmoodi-Meimand, H. Leakage Current Mechanisms and Leakage Reduction Techniques in Deep Sub-Micrometer CMOS Circuits. *Proc. IEEE* **2003,** *91*, 305–327.

Sahoo, K. C.; Oates, A. S. Dielectric Breakdown of Al_2O_3/HfO_2 Bi-Layer Gate Dielectric. *IEEE Trans. Device Reliab.* **2014,** *14*, 327–332.

Semenov, O.; Vassighi, A.; Sachdev, M. Leakage Current in Sub-Quarter Micron MOSFET: A Perspective on Stressed Delta I_{DDQ} testing. *J. Electron. Test. Theory Appl.* **2003,** *19*, 341–352.

Shauly, E. CMOS Leakage and Power Reduction in Transistors and Circuits: Process and Layout Considerations. *J. Low Power Electron. Appl.* **2012,** *2*, 1–29.

Sheu, B.; Scharfetter, D.; Ko, P.; Min-Chie, J. BSIM: Berkeley Short-Channel IGFET Model for MOS Transistors. *IEEE J. Solid-State Circuits* **1987,** *22*, 558–566.

Silvaco Inc. ATLAS User's Manual. Silvaco Inc., USA, 2013.

Sohn, C.; Sagong, H. C.; Jeong, E.; Choi, D.; Park, M.; Lee, J.; Kang, C.; Jammy, R.; Jeong, Y. Analysis of Abnormal Upturns in Capacitance Voltage Characteristics for MOS Devices with High-κdielectrics. *IEEE Electron Device Lett.* **2011,** *32*, 434–436.

Suzuki, K.; Tanaka, T.; Tosaka, Y.; Horie, H.; Arimoto, Y. Scaling Theory for Double-Gate SO1 MOSFET's. *IEEE Trans. Electron Devices* **1993,** *40*, 2326–2329.

Tarun, A.; Laniog, J.; Tan, J.; Cana, P. Junction Leakage Analysis using Scanning Capacitance Microscopy. *Proc. IEEE IPFA* **2003,** 213–216.

Taur, Y.; Ning, T. *Fundamentals of Modern VLSI Devices,* Cambridge University Press: New York, USA, 1998, pp 94–95.

Thompson, S.; Packan, P.; Bohr, M. MOS Scaling: Transistor Challenges for the 21st Century. *Intel Technol. J.* **1998,** *3*, 1–19.

Uejima, K.; Hase, T. Ultra Low Leakage Junction Engineering of Cell Transistor by Raised Source/Drain for Logic Compatible 28 nm Embedded DRAM. *Proc. IEEE VLSIT* **2011,** 170–171.

Vishnoi, R.; Kumar, M. 2-D Analytical Model for the Threshold Voltage of a Tunneling FET with Localized Charges. *IEEE Trans. Electron Devices* **2014,** *61*, 3054–3059.

Wang, T. Minimized Device Junction Leakage Current at Forward Bias Body and Applications for Low Voltage Quadruple-Stacked Common Gate Amplifier. *IEEE Trans. Electron Devices* **2014,** *61*, 1231–1236.

Wilk, G. D.; Wallace, R. M.; Anthony, J. M. High-k Dielectrics: Current Status and Material Properties Considerations. *J. Appl. Phys.* **2001,** *89*, 5243–5275.

Wu, Y.; Chen, L.; Lyu, R.; Wu, J.; Wu, M.; Lin, C. MOS Devices with High-k $(ZrO_2)5(La_2O_3)1-5$ Alloy as Gate Dielectric Formed by Depositing $ZrO_2/La_2O_3/ZrO_2$ Laminate and Annealing. *IEEE Trans. Nanotechnol.* **2012,** *11*, 483–491.

Wu, J.; Shen, Y.; Reinhardt, K.; Szu, H.; Dong, B. A Nanotechnology Enhancement to Moore's Law. *Appl. Comput. Soft Comput.* **2013,** *2013*, Article ID 426962, 13.

Yadav, M.; Akashe, S. New Technique for Reducing Sub-Threshold Leakage in SRAM. *Proc. IEEE ACCT* **2012,** 374–377.

Yeo, Y.; Lu, Q.; Ranade, P.; Takeuchi, H.; Yang, K.; Polishchuk, I.; King, T.; Hu, C.; Song, S.; Luan, H.; Kwong, D. Dual-Metal Gate CMOS Technology with Ultra Thin Silicon Nitride Gate Dielectric. *IEEE Electron Device Lett.* **2001,** *22*, 227–229.

Yu, B.; Wang, L.; Yuan, Y.; Asbeck, P. M.; Taur, Y. Scaling of Nanowire Transistors. *IEEE Trans. Electron Devices* **2008,** *55*, 2846–2858.

Yuan, Z.; Yu, Z. Short-Channel Effect Modeling of DG-FETs using Voltage-Doping Transformation Featuring FD/PD Modes. *IEEE Electron Device Lett.* **2009,** *30*, 1209–1211.

Zhu, W; Ma, T. P., Zafar, S.; Tamagawa, T. Charge Trapping in Ultra Thin Hafnium Oxide. *IEEE Electron Device Lett.* **2002,** *23*, 597–599.

CHAPTER 3

High-k Dielectric Materials: Structural Properties and Selection

P. SRI HARSHA, K. VENKATA SARAVANAN, and V. MADHURIMA

Department of Physics, School of Basic and Applied Sciences, Central University of Tamil nadu, Thiruvarur, Tamil nadu, India

*Corresponding author: E-mail: madhurima@cutn.ac.in

ABSTRACT

The role of high-k materials in the field of electronics is well known. Over the last decade or so, there has been an increase in the need for flexible electronics, thus necessitating a paradigm shift in the nature of high-k materials. Ceramics are being supplemented with polymers, other soft matter and composites. In this chapter, we discuss the theory of high-k materials and the properties of various classes of materials.

3.1 INTRODUCTION

In our semiconductor-driven technological society, miniaturization of circuits is a very important factor so as to fit in more and more components in a limited surface area. In order to achieve this miniaturization of electrical components, dielectric properties of the constituent devices such as capacitors, memories, sensors, resonators etc. (Tamala, 1999; Lau, 1994; Bai et al., 2000; Facchetti et al., 2005) play an important role. The current industry standard materials don't meet the criteria for miniaturization, for reasons discussed below. In addition, there is a need for wearable electronics, calling for elasticity of materials used. Thus, there is a need for novel soft matter that can replace current ceramic dielectric materials that are rigid.

High-k dielectrics are materials in which no steady current can flow through (Landau et al., 2013) as a result of which the static electric field in not zero as is the case in conductors. Dielectrics can be summarized as insulators that can be polarized by the application of electric field. The relationship between relative permittivity ε_r (generally k) and electric polarizability, P, is expressed by Clausius–Mossotti relation (eq 3.1):

$$\frac{\varepsilon-1}{\varepsilon+2}=\frac{1}{3\varepsilon_0}\Sigma N_j\alpha_j \tag{3.1}$$

where N_j is the concentration and α_j is the polarizability of the atom j, which is the sum of electronic, vibrational, and orientation polarization (Kittel, 2005). Ionic and inter-facial polarization are not accounted by Clausius–Mossotti relation, so in a sense it is important to understand polarization so as to understand dielectric constant since polarization is more universal in nature (Zhu, 2014).

Silicon dioxide has been used as a standard material in MOSFET (metal–oxide–semiconductor field-effect transistor) devices, which is a key component of microelectronic circuits. As the size of these MOSFET devices are further reduced, to push more and more devices into a limited area, two problems are seen (1) leakage currents and (2) reliability issues. This leakage current in MOSFETs is due to tunneling effect, where charge carriers flow through the gate barrier. Tunneling probability is inversely proportional to the thickness of the silicon dioxide gate, and hence this puts a limit on the size up to which the silicon dioxide layer can be reduced in thickness. The leakage current issue due to tunneling limits the thickness of SiO_2 layer in MOSFETs to around at 2–3 nm (Frank et al., 2001).

Another issue with SiO_2 is the reduced reliability as the thickness is reduced. When charge carriers flow through the gate they develop defects and when these defects reach a high enough value, the junction breaks down and the device fails. This issue of reliability is particularly pronounced when the thickness of the SiO_2 layer is reduced to nanometer ranges. Along with above-mentioned problems, there are few other issues pertaining to SiO_2 such as when SiO_2 is used a dielectric medium in between two conducting plates of a capacitor, and the relationship between capacitance C and the thickness of SiO_2 is given by eq 3.2:

$$C=\frac{A\varepsilon_r\varepsilon_0}{t_{ox}} \tag{3.2}$$

where t_{ox} is the thickness of the gate oxide layer, and A is the area of the capacitor. Thus, the capacitance can be increased for the same voltage by decreasing the thickness of SiO_2. However, in SiO_2 the thickness has reached the threshold value and any further reduction in thickness leads to quantum tunneling effects and increase in leakage current. Hence it's not possible to reduce the thickness of the gate oxide layer.

To tackle the above-mentioned issues, a material that has a higher *k* value has to be used (Houssa, 2003). Initially, efforts were made to make SiO_2 work by improving the *k* value by doping it with impurities like nitrides. These nitrides form oxynitrides and oxide–nitride stacks. Oxynitrides have better resistance toward a leakage current due to increased *k* value, and this increase is due to change in ionic polarizablity since oxygen is replaced by nitrides. Stacking causes the thin film to get thicker and hence reduces the penetration due to tunneling effect and provides better reliability. Generally, dielectric materials are divided into high-*k* and low-*k* materials in reference to the *k* value of SiO_2, which has a dielectric constant of 3.9. High-*k* materials are made of either ceramics or soft matter such as polymers.

3.2 CERAMIC HIGH-k MATERIALS

Ceramics are traditional dielectric materials that have high-k value, low breakdown strength and high dielectric loss. Ceramics have mechanical rigidity like high stiffness and strong thermal stability. Ceramic capacitors usually have high stability and are best applied in resonant circuits and they offer high volumetric efficiency for buffer, by-pass, and coupling applications (Herbert, 1985; Wersing, 1991; Rao et al., 2002). Ceramic-based field-effect transistors (FETs) are very difficult to manufacture since they require good deposition methods like aerosol deposition, sol–gel methods and other classic techniques like chemical vapor deposition (CVD), sputtering, evaporation etc. (Atsuki et al., 1995; Eranna et al., 2004; Levy et al., 2008) and high annealing temperatures.

Today, with electronic devices moving more toward free-form shapes, applicability of ceramics in electronic devices are restricted due to the higher density of ceramics and their poor flexibility. Thus, although ceramics were offering very good dielectric properties, the above limitations are a huge road block for current and future applications. Hence, there is a need for alternative materials to take care of these issues. The alternative to ceramic dielectrics is the polymer-based dielectrics.

3.3 POLYMER DIELECTRIC

Polymers are a class of soft matter, which cover a wide range of materials with a wide range of properties like flexibility, tunable thermal stability, molding capability, and volume and weight reduction. Polymers can be tuned for their dielectric properties. Their use in electronics as dielectrics is going up due to their tunability while retaining their useful properties such as flexibility, low weight, and low cost, leading to potential applications in conformal electronics. Of the various classes of polymers available, we will be considering here the electro-active polymers (EAPs). EAPs have been extensively used in recent years in a broad range of modern electrical applications such as flat panel display, microelectronic packaging, fiber optics, medical applications, fuel cells, high frequency transducers, artificial muscles, and other sensors and actuators (Bar-Cohen, 2001; Kim and Tadokoro, 2007; Tagarielli et al., 2012; Bar-Cohen, 2004).

In recent years, a new class of organic EPA has been developed to counter some of the shortfalls of general polymer EPAs like requirement of high electric field (Zhang et al., 1998; Xu et al., 2001) to produce a good response and hysteresis (Zhang et al., 2002). High-*k* polymers can be developed for use in electronics industry using various deposition techniques such as atomic layer deposition, chemical vapor deposition and pulsed laser deposition. Organic FETs also called OFET devices where the gate dielectric is organic in nature that has unique features like being ambipolar and opto-electronic (Coropceanu et al., 2007; Bredas et al., 2004; Gunes et al., 2007; Zaumseil and Sirringhaus, 2007). An ideal high-*k* material should not only have a high dielectric constant, but also have a very low dielectric loss and high breakdown strength.

Organic polymers are suitable materials as they usually have high breakdown strength and are easy to process. However, they have a low dielectric constant. Those with relatively high dielectric constants have high dielectric losses, particularly under high electric fields (Hwang et al., 2013).

Thus from above we can conclude that no single class of material is perfect for current application purposes and a new class of materials that combine the properties of both ceramics and polymers should be developed. Nanoparticles, due to their small size, can have a very large amount of active interacting surface areas that can be used as an advantage to create a new range of materials that have high dielectric constant, high breakdown strength, low dielectric losses, mechanical flexibility, and thermal stability.

3.3.1 Why High-k Polymers

As we have just seen, there is a need to create polymer composites that will have all the desirable properties and a precise *k* value for specific applications. As seen in the first section, *k* depends on polarization of material, which includes inter-facial, ionic, orientation, vibrational, and electronic polarization. One can achieve a very high *k* value by using more types of polarization in material, but each type of polarization has dielectric losses at different frequencies. In order to avoid loss dispersion, it is necessary to understand how polymers are polarized and how losses are mitigated by tuning the dielectrics to required standards.

3.3.2 Role of Polarization in k Value

Polarizations can be divided into two types: resonance and relaxation. Electronic and vibrational polarization have dielectric losses in frequency range between optical and infra-red ranges in general, hence are called resonance type polarization. The rest of orientational, ionic, and inter-facial polarizations are of relaxation type. The dispersion features of polarization are shown in Figure 3.1.

Depending on amorphous or crystalline type of dipoles and temperature, dipolar relaxation of polymers occupies a small region from 1 Hz and 100 MHz (Kao, 2004). Ionic polarization is a physical transport of ions in polymers. Polymeric electrolytes are made by designing ionic polarization, which is used to enhance the capacitive performance of gate dielectrics in OFET (Kim et al., 2013). This polarization aids in a very high energy storage performance, but still polymer electrolytes are not suitable for film capacitors because they have a high dielectric loss due to the transport of ion species over long distances. Therefore, the ionic polarization should be avoided for typical polymeric dielectrics by decreasing the concentration of impure ions during the polymerization processes. Ionic polarization takes place in a fraction of seconds or sometimes hours.

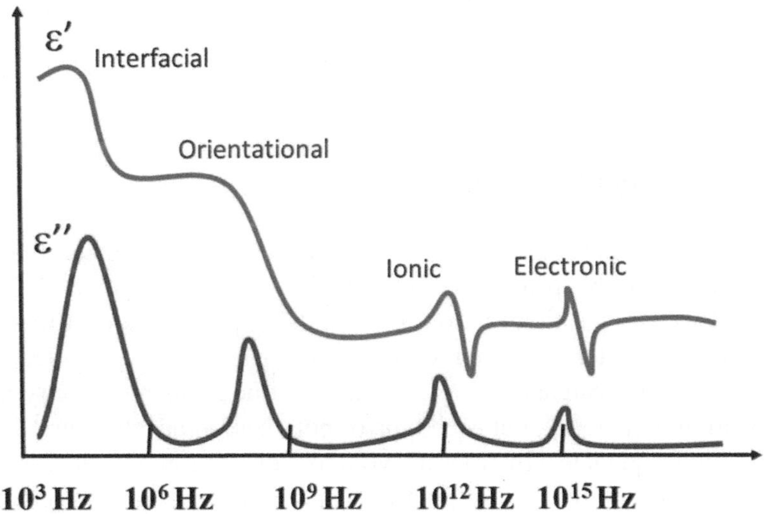

FIGURE 3.1 Real (ε') and imaginary (ε'') part of the dielectric constant as a function of frequency in a polymer having interfacial, orientational, ionic, and electronic polarization.

Inter-facial polarization or Maxwell–Wagner Sillars inter-facial polarization as the name suggests refers to the accumulation of charge carriers in a multi-layer polymer, when a current flows across the two-material interfaces, charges are accumulated in both materials involved in forming the interface, and these different dielectrics must have different relaxation time.

$$T = \frac{\varepsilon}{\sigma} \tag{3.3}$$

Where ε is the dielectric constant of a polymer, and σ is the conductivity. So, if there is a large difference in dielectric constant between filler and polymer matrix, it is highly desirable for interfacial polarization and can give rise to a high dielectric constant. It should be noted that it takes a large amount of time for these inter-facial charges to discharge (Sessler et al., 1998). Inter-facial polarization can be observed in semi-crystalline polymers, polymer blends, and nanocomposites. This huge difference in dielectric constant may result in electric field intensification in the polymer matrix especially on the particle surface (Ducharme, 2009). Depending on the temperature and type of dipoles (amorphous or crystalline), dipolar relaxation of polymers occupies a small region between 1 Hz and 100 MHz (Kao, 2004).

There are a few more sources that cause high losses in dielectrics like electronic conduction at high temperatures. Electronic conduction includes electrode-limited conduction mechanisms (Chiu, 2014). Electrode-limited conduction mechanisms include Schottky emission, Fowler–Nordheim, direct tunneling, and thermionic field emission. Bulk-limited conduction mechanisms include Poole–Frenkel emission, electron hopping, ohmic conduction, and space charge-limited conduction. Apart from these, electrochemical reactions must be nullified for dielectric applications. Usually, it is small polar molecules and ions that take part in electrochemical reactions, but whereas non-polar molecules and polymers do not readily undergo electrochemical reactions unless they are contaminated with small polar molecules and ions that readily undergo electrochemical reactions (Zhu, 2014). Therefore, it has to be ensured that polymer dielectrics should be free of moisture (polar molecules) and/or impurity ions. Therefore, one has to be very careful in choosing which type of polarization should be used for increasing the k value as each polarization has its own source of loss and sometime many loss mechanisms work in tandem.

3.3.3 Route to High-k Polymers

As discussed before, miniaturization of electronic circuits has caused the need to develop materials with high-k value, large breakdown strength, and very low dielectric losses. It has been known that high-k ceramic materials such as $BaTiO_3$ (BTO) can be fabricated into thin films by using chemical solution deposition and has a very high dielectric constant in the neighborhood of 2500 with low losses (Ihlefeld et al., 2005). Oxides of metals such as BTO are made by reactive sputtering by introducing the second element (oxygen) in gaseous form. However, other methods, such as atomic layer deposition (ALD) (Lin et al., 2002; Hausmann et al., 2002; Chang and Sawin, 2001; Gupta, 2009) and CVD, can be applied to deposit the high-k materials. The drawback however is that the substrate material cannot withstand temperatures (of the order of 900°C) that are required for the sintering process for BTO. There are high cost methods to create BTO nanoflims at room temperature like RF magnetron sputtering (Yao et al., 2012). Again the high cost and poor flexibility of ceramics are not attractive options.

Polymers are known for large-scale processability, high flexibility, high electric breakdown field, and are light weight, with the only drawback of low dielectric constant of less than 10. So, the modification of these polymers is a need and can be done through; modification of polymer chain structures and random composite approaches (Dang et al., 2012), to increase the dielectric constant. The best example for this is modification to polyvinylidene fluoride (PVDF) polymers resulting in dopamine functionalized PVDF through a refluxing method, with a dielectric constant of 32 and loss of 0.04 (Wang et al., 2010).

Composite technology is the best and easy way to produce materials with tunable dielectric properties, as it we can control the fillers shape, size, concentration etc. In subsequent sections, we will discuss various ways to achieve high-k in the following order:

- Ferroelectric ceramic/Polymer composites
- Conductive filler/Polymer composites
- All organic polymer composites
- Nanoparticle-based high-k composite
  - (a) Controlled dispersion
  - (b) Core–shell hybrid filler
  - (c) Metal nanoparticles
  - (d) Surface-modified metal nanoparticles
  - (e) High-*k* polymer matrix

3.4 FERROELECTRIC CERAMIC/POLYMER COMPOSITES

Ceramic particles with high-*k* can be infused into polymer matrices, which essentially have low-*k* to form dielectric composites. This represents one of most promising and exciting venues in this field (Dang et al., 2007; Dang et al., 2009). These composites are important to study because they not only provide high-*k* materials, but also possess the property where we can control the dielectric losses as seen in Figure 3.2. These composites have the best of both polymers and ceramics, i.e. they are flexible, easily made, have high dielectric permittivity, and have high breakdown strength. Though ceramics have very high dielectric constant, it's important to control the amount as large percentage of the ceramics will cause percolation.

The maximum amount of filler concentration is around 60% after which the percolation point is reached. Dielectric constant of polymers can be increased using ferroelectric metal oxides as filler, such as TiO_2, ZrO_2, BTO, and $CaCu_3Ti_4O_{12}$ (CCTO) (Dang et al., 2009; Balasubramanian et al., 2010; Dang et al., 2011; Zou et al., 2011; Tuncer et al., 2008; Yang et al., 2011; Prakash and Varma, 2007; Amaral et al., 2008). For these composites to produce a very high-*k* value, a very high loading of inorganic nanoparticles to about 3040 % is required, and this high percentage of filler causes many problems like loss of mechanical strength and increased porosity that decreases dielectric properties of composites.

The problem of loss of strength and dielectric properties due to filler can be overcome by using rod shaped nanoparticles, for example, electrospun BTO nanofibers with a large aspect ratio as dielectric fillers in poly(vinylidene fluoride-trifluoroethylene) (PVDF-TrFE)-based nanocomposites (Hu et al., 2013). Composites formed by nanotubes show a dielectric constant of 30 with a reduced filler loading (10.8 vol%). Furthermore, BTO nanotubes were seen to be more efficient than rods as they have very high aspect ratio yet less mass density (Liu et al., 2014).

Till recently, 2D high-k nanoplatelets for enhancing the dielectric constant of polymers was non-existent, but now 2D carbon nanomaterials with large aspect ratios—graphene nanosheets (GNs)—have become 2D materials to prepare high-*k* nanocomposites (Dang et al., 2016). The GNs and polymer composite owing to their large surface area give a very high-*k* nanocomposite. This large surface area generates a low percolation threshold. The use of GNs in polymer matrix helps in producing high-*k* as many microcapacitors are formed in composites. Apart from generating high-*k* values, it also helps in the preparation process and improves the intrinsic properties of the polymer. As in the case of PVDF that is a popular material as a polymer composite due to its piezoelectric effects and dielectric properties. When GNs are introduced as filler, they have a very good effect on the matrix that acts as a nucleation agent in PVDF and helps in the generation of β-phase of PVDF that has better dielectric properties (Gregorio and Ueno, 1999). It has been verified that GNs and PVDF polymer nanocomposites with a loading of 0.1 wt% GNs exhibited almost all β-phases, and a high permittivity of 41 at 103 Hz (Huang et al., 2014).

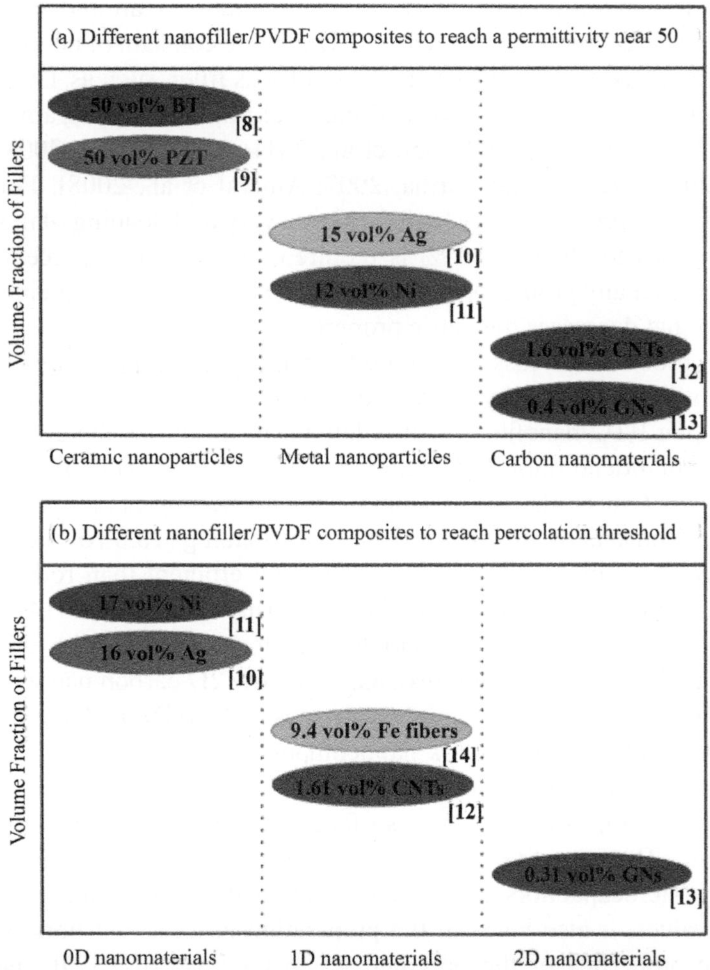

FIGURE 3.2 (a) Scheme of different nanofillers/PVDF composites, in which the nanofillers were categorized as ceramic nanoparticles, metal nanoparticles, and carbon nanomaterials. This scheme shows the corresponding volume fractions of nanofillers in composites to get a similar permittivity of near 50. (b) Scheme of different dimensional nanomaterials/PVDF composites with the corresponding percolation threshold, in which the nanofillers were categorized as 0D nanomaterials, 1D nanomaterials, and 2D nanomaterials. This scheme shows the corresponding volume fractions of nanofillers in composites to reach a percolation threshold (Adapted from Dang et al., 2016).

Although a high-*k* can be achieved in polymer composites by loading various high-*k* fillers, there are still many technological and scientific challenges in realizing high-performance nanocomposites. The usual problems include method to achieve uniform dispersion of inorganic nanoparticles, achieving desirable inter-facial polarization. Usually, coupling agents like silane are used as coupling agent in BTO/polyvinylidene fluoride composites (Dang et al., 2006), and surfactants (Ramesh et al., 2003) are used in order to modify the properties of nanoparticles that can increase the dielectric constant of the polymer composite. Sometimes, these modifications act as a negative influence that can lead to a decrease in the dielectric strength of the polymer matrix. They can sometime cause excessive agglomeration of ceramic fillers as the inorganic filler and polymer matrix are not compatible, which unfortunately can cause an increase in the dielectric loss and reduction of the breakdown field (Huang and Zhang, 2004).

3.5 CONDUCTING NANOPARTICLE FILLER–POLYMER COMPOSITES

Metal particles that are conductors are not dielectric but have a negative *k* value. When these particles used in polymer nanocomposites, they show a considerably higher dielectric constant than ceramic dielectrics. As seen in the above section, we can even achieve an even higher dielectric constant if we can have particle loading just a little lower than that of the percolation threshold (Yuan et al., 2011). With increasing particle concentration in a polymer matrix, these particles can form numerous micro-capacitors that would add up to a very high capacitance of the nanocomposite, which causes the composite to have a very high dielectric constant as seen in Figure 3.3.

As compared to the spherical conductive particle, 1D carbon materials with large aspect ratios can lead to a much lower percolation level. Yuan et al. (2011) has used carbon nanotubes (CNT) instead of metal nanoparticles to create high-*k* polymer nanocomposites. These CNT-based composites were prepared by melt mixing multiwalled carbon nanotubes (MWNT) within PVDF host polymer. Similar to nanoparticles, as we approach percolation, microcapacitors are formed with adjacent nanotubes as electrodes and polymer matrix as dielectric. The use of CNT has the advantage of high surface area of CNT, which helps in increase polarizability. CNTs also help in increasing the inter-facial polarization inside the micro capacitors owing to the formation of donor–acceptor complexes.

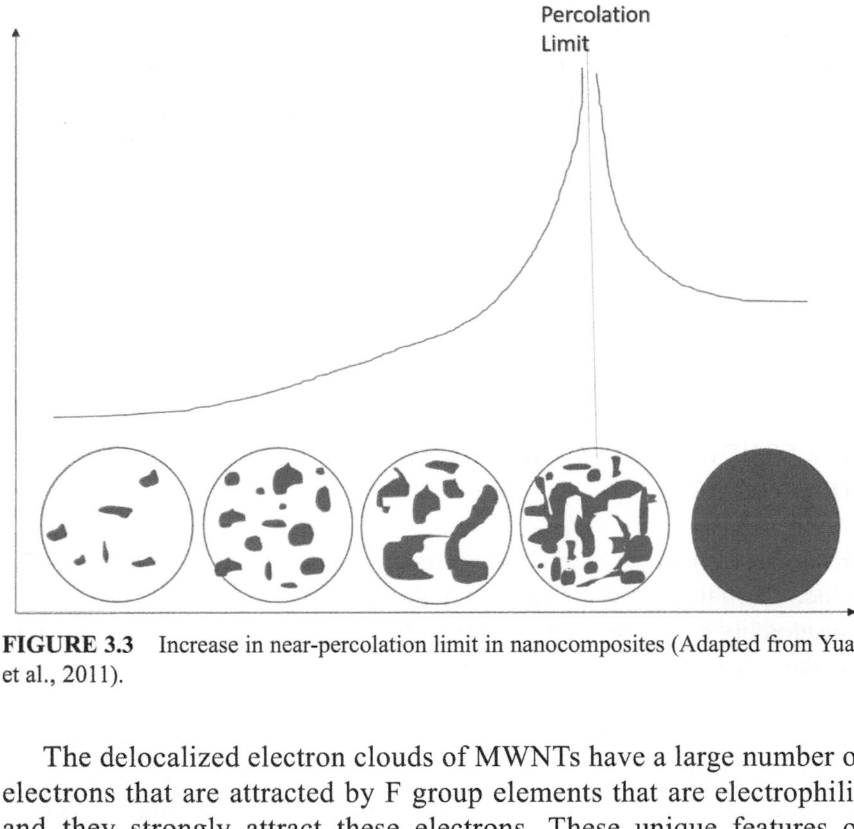

FIGURE 3.3 Increase in near-percolation limit in nanocomposites (Adapted from Yuan et al., 2011).

The delocalized electron clouds of MWNTs have a large number of electrons that are attracted by F group elements that are electrophilic and they strongly attract these electrons. These unique features of individual CNT networks give rise to a very high dielectric constant of 3800 at 1 Hz (Yuan et al., 2011). In the above section, another possible method for high dielectric is percolating the polymer matrix with 2D conductive filler with a disk or plate shape. And this large k values in conducting particle/polymer composites is due to inter-facial polarization. These composite dielectrics are not without their own demerits as charges inside the composite have a high possibility of being delocalized through tunneling effect or by ohmic conduction (Huang and Zhi, 2016). These both processes lead to a dissipation factor. Even though this procedure of composites provides us with high dielectric constant, there is also a very high loss factor that needs to be mitigated. There is still a search going on for finding a strategy to avoid possible loss. One of the most promising one is the use of a layer of material that acts as a barrier between two particles in the matrix and this can reduce the tunneling effect quite considerably, which results in a reduced dielectric loss (Zhou et al., 2011; Wu et al., 2012; Wu et al., 2011). One such use of barrier to reduce the tunneling effect is an organic polysulfone (PSF) dielectric shell on MWNTs by electrospinning method (Zhang et al., 2012). The PSF shell here serves a sole purpose as a barrier to prevent MWNTs from direct contact and also as a polymer matrix after fusion by hot pressing. The resultant MWNT/PSF composites showed a high dielectric constant when the nanotube content reached 25 vol%, while there was only a slight change in loss tangent from 0.02 to 0.05 (Huang and Zhi, 2016).

Other method to reduce a dielectric loss is by the use of heterogeneous spatial distribution of conducting particles (Wang et al., 2013; Wu et al., 2011). Recently, Wu et al. (2011) have developed MWNT/cyanate ester (CE) composite with a gradient dielectric constant via the gradient distribution of MWNT along the direction of the thickness (Huang and Zhi, 2016). The advantage here is that CE-rich region has the ability to break the conducting path along the direction of electric field, this break in conduction results in a lower loss. Yuan et al. (2014) successfully increased the dielectric constant by vertically aligned CNT arrays as filler (Yuan et al., 2014). When the polymer matrix is penetrated into these CNTs, they behave like miniature microcapacitors. This method of vertical-CNT is very efficient, if we are able to penetrate the arrays into the polymer matrix without disturbing the structure. Most important part is that these improvements in dielectric constant are possible to be achieved at very low loading, showing a much higher potential than that of the percolated particles nanocomposites. The most important aspect of all these techniques is to have an optimal balance in the percolation of conducting particles in the composite. New progress is happening everyday where new techniques are being developed to have better dielectric constant with low losses. Nanoscale 3D printing and similar technology might be of great value in creating an array of organized structures that have low losses and can be tuned and constructed according to our needs.

3.6 ALL ORGANIC POLYMER COMPOSITES

All organic polymer composites gained attention and got a requirement to achieve high dielectric constant that is due to its requirements in electroactive polymers (EPA). These EPA have the ability to convert charge and voltage to mechanical force and vice versa; the response of these materials can be controlled by applied electric filed. The problem arises as the field required to extract a response is very high and thus needs a very high field (Bar-Cohen, 2001; Xu et al., 2002). The energy generated from EPA cannot exceed the input energy (Huang et al., 2004). The major breakthrough in the field was by Zhang et al. (2002) where they created a two-component composite that can exhibit high elastic energy densities with a very little applied electric field. This organic composite is fabricated from a filler material with very high dielectric constant dispersed in an electrostrictive polymer matrix. The result is a material composite with high dielectric constant (Zhang et al., 2002).

The procedure of using composites by addition of high dielectric particles to increase the dielectric constant of the matrix is not new. However, these fillers (frequently ceramics or metal particles) also possess very high elastic moduli than that of polymers. The resulting composite also show an elastic modulus much higher than that of the polymer matrix that causes a loss of flexibility. Zhang et al. (2002) selected a metallophthalocyanine (MtPc) oligomer, copperphthalocyanine (CuPc), as the filler of high dielectric constant higher than 10000. A dielectric constant of about 105 was reported in CuPc oligomers. The large dielectric constant of this composite is due to electron delocalization within MtPc molecules (Nalwa, 1999). The easy displacement of the electrons under electric fields from the conjugated π-bonds within the molecule causes a higher dielectric response. Also, the weak Van der Waals intermolecular forces cause the elastic modulus around the same value that of the matrix. MtPc solids show a high dielectric loss because of the long-range intermolecular hopping of electrons (Gould, 1996). The polymer matrix forms insulation layers to reduce the dielectric loss of the filler. The resulting composite exhibits almost the same elastic modulus as the polymer matrix and retains its flexibility (Zhang et al., 2006).

One of the widely used ways to increase the dielectric constant is through the use of high dielectric organic composites, where organic molecules are blended with polymer to make a composite with significant increase in the dielectric constant (Zhang et al., 2006). Another approach is by a percolative composite one, where instead of metal fillers to increase the dielectric constant a conductive polymer is used. One such method was reported by Huang et al. where they used polyaniline (PANI), as the conductive filler due to its relatively low elastic modulus in comparison to metal particles. It was reported to show a dielectric constant of 2000 (at 100 Hz) for a composite containing 23 vol% of PANI particulates (Huang et al., 2003). At 100 Hz, the dielectric constant of a composite with only 12.7% of c-PANI filler is only 10 times higher than the matrix but with increase in concentration to $f = 23\%$, the dielectric constant rises nearly 2000 times that of the matrix. However, any further increase in the percentage of the filler, the dielectric loss increases and the breakdown field reduces quite rapidly, due to the approach of percolation (Pecharroman et al., 2001; Efros and Shklovskii, 1976).

The above-mentioned method to achieve an organic dielectric constant has a drawback in the form of the volume fraction of the conductive polymer, as it must be very close to the percolation threshold, f_c, in order to reach the high dielectric constant, this is because of the predicted relation of conductive fillers in an insulation matrix for a composite material (Efros and Shklovskii, 1976; Nan, 1993).

$$K_c = k_m \left\{ \frac{f_c - f}{f} \right\}^{-q} \tag{3.4}$$

In eq 3.4, if f approaches f_c, the breakdown field becomes very low and this has to be avoided for a robust composite. In a result published by Shen et al. (2007), they have shown that the carbonaceous shells coating the silver cores determine the dielectric behavior of the percolative nanocomposites having Ag cores coated by organic carbonaceous shells as fillers and epoxy as the polymer matrix. The insulating shells around the conductive metal fillers isolate the silver metallic cores from each other, this separation allows the formation of nanocapacitor networks in the polymer matrix and results in higher and stable dielectric constants even when $f > f_c$. Tunable dielectric constants as discussed above by fine tuning the material properties have been achieved by adjusting the thickness of the organic insulating shells, and their dielectric constants are almost independent of the frequency over a very wide frequency range (Shen et al., 2007). These metal nanoparticle-based organic polymer dielectrics have the unique property that they form cross-linked structures in reducing the charge trapping sites and thus resulting a reduction of hysteresis without the loss of dielectric constant. This was shown by Beaulieu et al. (2013) in their research publication where they used ZrO_2 nanoparticles in cyano-ethyl pullulan (CYELP), which is a high-*k* polymer. This combination of high-*k* nanoparticle and high-*k* polymer was used to construct an OFET with increased *k* value. The results show the dielectric constant around 15.6–21 based on the ratios of the composition materials at 100 Hz (Beaulieu et al., 2013).

3.7 CORE–SHELL STRUCTURED HIGH-k POLYMER NANOCOMPOSITES

For enhancing properties of nanoparticles inside the nanocomposites, significant efforts have recently been devoted to the design and synthesis of core–shell nanoparticles. Best example to consider is the case of BTO for which case there are some popular methods to create high-*k* polymer nanocomposites using core–shell structure.

They are:

(i) Direct use of core–shell nanoparticles prepared by grafting from
(ii) Direct use of core–shell nanoparticles prepared by grafting to
(iii) Using core–shell organic–inorganic nanoparticles as fillers
(iv) Using other types of core–shell nanoparticles as fillers

We will discuss the two major practices in developing core–shell particles: grafting to and grafting from. In the discussion below, we can go through these methods briefly to understand that high-*k* dielectrics are fine-tuning between the properties of its composition materials where the dielectric constant, dielectric loss, and breakdown strength are adjusted in order to achieve a desirable composite material.

3.7.1 Grafting From

In the first method "grafting from", core–shell nanoparticles is created by numerous initiating sites on the surface of particles. This method relies on the polymerization of monomers on the initiator sites on the surface of nanoparticles. One of the prime examples is done by Xie et al. (2011), in which they have shown that the surface of the core–shell can be used as the matrix and this eliminates the need to add further polymers to create the matrix. Method used by Xie et al. (2011) allows for the preparation of thin, defect-free nanocomposites bearing high filler loadings (Huang and Zhi, 2016) in which core–shell BTO/poly(methyl methacrylate) (PMMA) nanocomposites via in situ atom transfer radical polymerization (ATRP) of methyl methacrylate (MMA) from the surface of BTO nanoparticles with the help of saline monolayers as initiators. Here, each material offers a different use like BTO core gives a very high-*k* (15 at 1 kHz), while the PMMA shell allows the material to have a good dispersibility and inherent low loss of the host polymer (0.037 at 1 kHz) by retarding charge carriers movement in composites (Huang and Zhi, 2016).

These above discussed core–shell composites have very less loss in comparison with that of composites with polymerceramic nanocomposites. The use of surface initiators for growth of polymer is demonstrated by using phosphonic acid as an initiator to grow polystyrene and PMMA from BTO nanoparticles with activators regenerated by electron transfer (Paniagua et al., 2014). The process is clearly explained in Huang and Zhi (2016). When the polymer matrix is attached covalently to the surface of the nanoparticles, a three-constituent system is created with a dielectric constant of nearly 11.4 at 1 kHz for 22 vol% BTO particle, while the loss tangent remains same.

One of the better methods to have a successful grafting from technique for core–shell particles is by controlled/living radical polymerization, such as ATRP and reversible addition-fragmentation chain transfer (RAFT) polymerization (Huang and Jiang, 2015). This method has some inherent advantages that make it worthwhile to study it further and fine-tune its properties. One of these advantages is that since the shell layer acts as a sheath and prevents the nanoparticle aggregation, these particles can be directly used as composites as the surface polymer acts as the matrix for the composite. This is one of the better composites as it is devoid of pores and defects in general. Due to the presence of the initiator of sites on the surface, there is a very strong bonding of the polymer on the surface of the nanoparticle thus providing a strong matrix interface. As initiator sites are controlled, there is a way to adjust and tune the amount of nanoparticles in the matrix and thus a controlled dielectric constant.

Another material made with the use of grafting from technique is biaxially oriented polypropylene (BOPP) film based capacitors that have a very unique combination of properties like a low energy loss, high breakdown strength, and low capacity loss under high frequencies. However, the low dielectric constant of PP limits the utilization of BOPP. A new method was developed to prepare PP-based nanocomposite that have low dielectric loss: it is made by coating the nanoparticles with methylaluminoxane, MAO, which facilities the formation of the Al–O bonding that is later exposed to react with metallocene catalyst that after in situ propylene polymerization gives isotactic PP (iso PP) nanocomposites. The PP nanocomposites had higher dielectric constant and larger than before breakdown strength and thus exhibited significantly enhanced energy storage capacity compared with pure PP. Grafting from technique has numerous advantages varying from huge scale production, activated catalyst centers for increased hydrostatic pressure, minimal dielectric mismatch of matrix, and nanoparticles. Due to the tunneling effect, the conductive nanoparticle-filled composite has a low breakdown strength, but in the above case of PP where high breakdown strength observed have to be attributed to the aluminum surface coating as this will inhibit the interparticle tunneling effect.

One common issue with the technique of "grafting from" is that during the synthesis there is a heterogeneous nucleation where a few particles are core less; this was overcome by Niitsoo et al. (2011), they report the formation of $Ag@SiO_2$ core–shell particles (Niitsoo et al., 2011), where there were no coreless silica particles. In ethanol, it depends on the available silver surface area whether homogeneous nucleation of silica on silver is achieved. They found that in methanol and 1-butanol, coreshell particles did not form. This demonstrates the significance of controlling the tetra-alkoxysilane hydrolysis rate (Brinker, 1988) when growing silica shells on silver nanoparticles.

3.7.2 *Grafting To*

In this method, core–shell nanoparticles are prepared by grafting the already formed polymer chains onto the surface of nanoparticles by the reaction between ends of the polymer chain and the surface of the nanoparticle. This technique in comparison with "grafting from" gives a higher freedom in the form of control in molecular composition and the molecular weight of the polymer chains with regard to the desired quality and performance requirement of the matrix.

"Grafting to" technique take advantage of click chemistry as it has a very high efficiency, good reactive conditions. One such instance of "grafting to" strategy is the use of click chemistry to attach PS to TiO_2 nanorods that showed a dielectric constant of 6.4 at 1 kHz and a loss of 0.625, this process is a rather noted one as it provide a very high degree of tunability of the polymer and thus that of nanocomposite (Tchoul et al., 2010). This dielectric can be used as a thin film transistor based on the organic semiconductor to achieve higher carrier mobility and low leakage current, with only disadvantage of catalyst removal as CuBr cannot be easily removed; however, thiolene reaction is highly efficient and have no by-products and no requirement of transition metal catalyst and thus thiolene click chemistry is a much preferable method for the preparation of core–shell nanoparticles.

Thiol-terminated PS or PMMA macromolecular chains with different molecular weights are prepared by RAFT polymerization, and core–shell polymer $BaTiO_3$ nanoparticles were fabricated by grafting the macromolecular chains on the surface of vinyl-functionalized $BaTiO_3$ nanoparticles via the thiolene click reaction. These nanocomposites have a higher dielectric constant; however, the dielectric loss was still as low and same as that of the pure polymer. This goes to prove that the dielectric loss of the core–shell polymer of $BaTiO_3$ nanocomposites is dependent on the molecular weight of the polymer chains and the grafting density of the core–shell structured nanoparticles (Dang et al., 2016). Not only click chemistry reaction but any other linking reaction between organic molecular chains and nanoparticle surfaces can also be used to prepare core–shell structured high-*k* nanocomposites.

3.8 CONTROLLED DISPERSION

Dispersion of nanoparticle is such an important part in creating a high dielectric material as when nanoparticles are used to increase the dielectric constant, agglomeration of nanoparticles will cause the material to have undesired electric properties. To have a desired uniform dispersion, numerous methods are employed like the addition of surfactant or dispersant such as phosphate esters (Bhattacharya and Tummala, 2000).

Chemical modification of nanoparticles is another approach for the dispersion of nanoparticles as well. Also Nariman Yousefi et al. (2014) have shown that low percolation threshold of 0.12 vol% is achieved in the rGO/epoxy system due to uniform dispersion of monolayer graphene sheets with very high aspect ratios (>30,000) (Nariman Yousefi et al., 2014). This uniform dispersion at above filler content causes the material to have a well-structured alignment and formation of conductive network along this preferred direction, this induces anisotropy in electrical and mechanical properties. This uniformly distributed rGOs have very high dielectric constants of 14,000 with just 3 wt% of rGO at 1 kHz due to the charge accumulation at the above-mentioned aligned conductive filler/insulating polymer interface, which is explained by the Maxwell–Wagner–Sillars polarization principle (Schonhals and Kremer, 2003).

Jiongxin Lu et al. (2006) have also reported that low dielectric losses in the in situ formed silver (Ag) incorporated carbon black (CB)/polymer composites. Addition of well dispersed silver nanoparticles has led to a very high dielectric constant and low losses. This was explained due to the Coulomb blockade effect of the Ag nanoparticles without which the composite has a high dissipation factor with just CB, which will give rise to a high dielectric losses.

Michael Bell et al. (2017) show that the filler dispersion and surface chemistry play a major role on the property enhancement in silica-epoxy nanocomposites (Michael Bell et al., 2017). There are many publications that show for the improvement in dielectric properties owing to the better dispersion of nanoparticle onto the matrix (Yang et al., 2013; Lu et al., 2006; Wan et al., 2016).

3.9 METAL NANOPARTICLES, SURFACE MODIFIED METAL FILLER, AND HIGH-k POLYMER MATRIX

The dielectric constant of the percolative composite is increased when filler loading approaches the percolation threshold (Pecharroman and Moya, 2000). This percolative method is advantageous than that of the traditional method of ferroelectric ceramic particles. Lu et al. (2008) have developed silver/polymer nanocomposite as a high matrix in which these metal nanoparticles were dispersed. This system showed a lesser dielectric loss (Lu et al., 2008). To reduce the dielectric loss, there have been many attempts such as creating a three phase nanocomposites, core–shell nanoparticles, and semiconductor fillers (Lu et al., 2006; Lu et al., 2008; Choi et al., 2006; Dang et al., 2002; Rao et al., 2005). As the percolation levels are reached, the metal nanoparticles form conductive paths within the matrix and cause the dissipation to increase and the dielectric constant to fall. In order to reach higher levels of percolation and increase the dielectric constant further, Li et al. (2016) have found a novel way in which a metal nanoparticle is coated with an insulating shell that prevents the contact of metal particles and thus avoiding the dissipation (Li et al., 2016).

Though metal nanoparticles are very useful in producing high dielectric constant but are limited due to its percolation limits, the physical properties of composite material change when filler particles form a percolating network through the composite. Nan et al. (2010) by use of electric conductivity and dielectric properties as examples reviewed the recent studies on the physical properties of composites near percolation. Properties of the interface between fillers and matrix on electric and dielectric properties near percolation that form an important concept of understanding the modifications required to achieve high-*k* can be understood in the discussion (Nan et al., 2010). This review concludes with an outlook on the future possibilities and scientific challenges in the field. As said before, this problem of reaching the percolation limit can be bypassed by modifying the surface of metal particles. This modification to the surface of the metal nanoparticles can help in the prevention of the formation of conductive paths between the particles (Qi et al., 2005).

KEYWORDS

- **high-*k* materials**
- **nanocomposites**
- **polymer**
- **nanoparticles**
- **ferroelectric**
- **ceramic**
- **dispersion**

REFERENCES

Amaral, F.; Rubinger, C.; Henry, F.; Costa, L.; Valente, M.; Barros-Timmons, A. Dielectric properties of polystyrene cc to composite. *J. Non-Crystalline Solids* **2008,** *354* (47), 5321–5322.

Atsuki, T.; Soyama, N.; Yonezawa, T.; Ogi, K. Preparation of bi-based ferroelectric thin films by sol-gel method. *Japan. J. Appl. Phys.* **1995,** *34* (9S), 5096.

Bai, Y.; Cheng, Z. Y.; Bharti, V.; Xu, H.; Zhang, Q. High dielectric constant ceramic-powder polymer composites. *Appl. Phys. Lett.* **2000,** *76* (25), 3804–3806.

Balasubramanian, B.; Kraemer, K. L.; Reding, N. A.; Skomski, R.; Ducharme, S.; Sellmyer, D. J. Synthesis of monodisperse TiO_2-paraffin core-shell nanoparticles for improved dielectric properties. *ACS Nano* **2010,** *4* (4), 1893–1900.

Bar-Cohen, Y. *Electroactive polymers as artificial muscles: Reality and Challenges.* SPIE Press: USA, 2001.

Bar-Cohen, Y. Transition of eap material from novelty to practical applications are we there yet? *Smart Struct. Mat.: Elect. Polymer Actuators Devices* **2001,** *4329*, 28.

Bar-Cohen, Y. *Electroactive polymer (EAP) actuators as artificial muscles: reality, potential, and challenges.* SPIE Press: USA, 2004; pp 136.

Beaulieu, M.; Baral, J.; Hendricks, N.; Tang, Y.; Briseno, A.; Watkins, J. Solution processable high dielectric constant nanocomposites based on ZrO_2 nanoparticles for flexible organic transistors. *ACS Appl. Mater. Interfaces* **2013,** *5* (24), 13096–13103.

Bell, M.; Krentz, T.; Nelson, J. K.; Schadler, L.; Wu, K.; Breneman, C.; Zhao, S.; Hillborg, H.; Benicewicz, B. Investigation of dielectric breakdown in silica-epoxy nanocomposites using designed interfaces. *J. Colloid Interface Sci.* **2017,** *495*, 130–139.

Bhattacharya, S. K.; Tummala, R. R. Next generation integral passives: materials, processes, and integration of resistors and capacitors on pwb substrates. *J. Mater. Sci.: Mater. Electron.* **2000,** *11* (3), 253–268.

Bredas, J. L.; Beljonne, D.; Coropceanu, V.; Cornil, J. Charge-transfer and energy-transfer processes in-conjugated oligomers and polymers: a molecular picture. *Chem. Rev.* **2004,** *104* (11), 4971–5004.

Brinker, C. Hydrolysis and condensation of silicates: effects on structure. *J. Non-Crystalline Solids* **1988,** 100, 31–50.

Chang, J. P. and Sawin, H. H. Molecular-beam study of the plasma-surface kinetics of silicon dioxide and photoresist etching with chlorine. *J. Vacuum Sci. Technol. B: Microelectron. Nanometer Struct. Process. Meas. Phenom.* **2001,** *19* (4), 1319–1327.

Chiu, F. C. A review on conduction mechanisms in dielectric films. *Adv. Mater. Sci. Eng.* **2014,** *2014*, 1–18.

Choi, H.-W.; Heo, Y.-W.; Lee, J.-H.; Kim, J.-J.; Lee, H.-Y.; Park, E.-T.; Chung, Y.-K. Effects of $BaTiO_3$ on dielectric behavior of $BaTiO_3$ nipolymethyl methacrylate composites. *Appl. Phys. Lett.* **2006,** *89* (13), 132910.

Coropceanu, V.; Cornil, J.; Filho, D. A.; Olivier, Y.; Silbey, R.; Bredas, J. L. Charge transport in organic semiconductors. *Chem. Rev.* **2007,** *107* (4), 926–952.

Dang, Z.-M; Shen, Y.; Nan, C.-W. Dielectric behavior of three-phase percolative nibatio 3/ polyvinylidene fluoride composites. *Appl. Phys. Lett.* **2002,** *81* (25), 4814–4816.

Dang, Z.-M.; Wang, H.-Y.; Xu, H.-P. Influence of silane coupling agent on morphology and dielectric property in $BaTiO_3$/polyvinylidene fluoride composites. *Appl. Phys. Lett.* **2006,** *89* (11), 112902.

Dang, Z.-M.; Xu, H.-P.; Wang, H.-Y. Significantly enhanced low-frequency dielectric permittivity in the BaTiO3/poly(vinylidene fluoride) nanocomposite. *Appl. Phys. Lett.* **2007,** *90* (1), 012901.

Dang, Z.-M.; Zhou, T.; Yao, S.-H.; Yuan, J.-K.; Zha, J.-W.; Song, H.-T.; Li, J.-Y.; Chen, Q.; Yang, W.-T.; Bai, J. Advanced calcium copper titanate/polyimide functional hybrid films with high dielectric permittivity. *Adv. Mater.* **2009,** *21* (20), 2077–2082.

Dang, Z.-M.; Xia, Y.-J.; Zha, J.-W.; Yuan, J.-K.; Bai, J. Preparation and dielectric properties of surface modified TiO_2/silicone rubber nanocomposites. *Mater. Lett.* **2011,** *65* (23), 3430–3432.

Dang, Z.-M.; Yuan, J.-K.; Zha, J.-W.; Zhou, T.; Li, S.-T.; Hu, G.-H. Fundamentals, processes and applications of high permittivity polymermatrix composites. *Progress Mater. Sci.* **2012,** *57* (4), 660–723.

Dang, Z.-M.; Zheng, M.-S.; Zha, J.-W. 1D/2D Carbon Nanomaterial-Polymer Dielectric Composites with High Permittivity for Power Energy Storage Applications. *Small* **2016,** *12*, 1688–1701.

Ducharme, S. An inside-out approach to storing electrostatic energy. *ACS Nano* **2009,** *3* (9), 2447–2450.

Efros, A.; Shklovskii, B. Critical behaviour of conductivity and dielectric constant near the metal–non-metal transition threshold. *Phys. Status Solidi (b)* **1976,** *76* (2), 475–485.

Eranna, G.; Joshi, B.; Runthala, D.; Gupta, R. Oxide materials for development of integrated gas sensorsa comprehensive review. *Critic. Rev. Solid State Mater. Sci.* **2004,** *29* (3-4), 111188.

Facchetti, A.; Yoon, M. H.; Marks, T. J. Gate dielectrics for organic field-effect transistors: New opportunities for organic electronics. *Adv. Mater.* **2005,** *17* (14), 17051725.

Frank, D. J.; Dennard, R. H.; Nowak, E.; Solomon, P. M.; Taur, Y.; Wong, H. Device scaling limits of simosfets and their application dependencies. *Proc. IEEE* **2001,** *89* (3), 259288.

Gerhard-Multhaupt, R. *Electrets*. Laplacian Press: Morgan Hill, CA, 1998.

Gould, R. Structure and electrical conduction properties of phthalocyanine thin films. *Coord. Chem. Rev.* **1996,** *156*, 237–274.

Gregorio, R.; Ueno, E. Effect of crystalline phase, orientation and temperature on the dielectric properties of poly(vinylidene fluoride) (pvdf). *J. Mater. Sci.* **1999,** *34* (18), 4489–4500.

Gunes, S.; Neugebauer, H.; Sariciftci, N. S. Conjugated polymer-based organic solar cells. *Chem. Rev.* **2007,** *107* (4), 1324–1338.

Gupta, T. *Copper Interconnect Technology*. Springer: New York, USA, 2009.

Hausmann, D. M.; Kim, E.; Becker, J.; Gordon, R. G. Atomic layer deposition of hafnium and zirconium oxides using metal amide precursors. *Chem. Mater.* **2002,** *14* (10), 4350–4358.

Herbert, J. *Ceramic Dielectrics and Capacitors*. CRC Press: USA, 1985.

Houssa, M. *High-K Gate Dielectrics*. CRC Press: USA, 2003.

Hu, P.; Song, Y.; Liu, H.; Shen, Y.; Lin, Y.; Nan, C.-W. Largely enhanced energy density in flexible p (vdf-trfe) nanocomposites by surface-modified electrospunbasrtio 3 fibers. *J. Mater. Chem. A* **2013,** *1* (5), 1688–1693.

Huang, X.; Jiang, P. Core-shell structured high-k polymer nanocomposites for energy storage and dielectric applications. *Adv. Mater.* **2015,** *27* (3), 546–554.

Huang C.; Zhang, Q. Enhanced dielectric and electromechanical responses in high dielectric constant all-polymer percolative composites. *Adv. Funct. Mat.* **2004,** *14* (5), 501–506.

Huang, X.; Zhi, C. *Polymer Nanocomposites: Electrical and Thermal Properties.* Springer International Publishing: USA, 2016.

Huang, C.; Zhang, Q.; Su, J. High-dielectric-constant all-polymer percolative composites. *Appl. Phys. Lett.* **2003,** *82* (20), 3502–3504.

Huang, C.; Zhang, Q.; Debotton, G.; Bhattacharya, K. All organic dielectric-percolative three-component composite materials with high electromechanical response. *Appl. Phys. Lett.* **2004,** *84* (22), 4391–4393.

Huang, L.; Lu, C.; Wang, F.; Wang, L. Preparation of pvdf/graphene ferroelectric composite films by in situ reduction with hydrobromic acids and their properties. *RSC Adv.* **2014,** *85*, 45220–45229.

Hwang, S. K.; Bae, I.; Cho, S. M.; Kim, R. H.; Jung, H. J.; Park, C. High performance multi-level non-volatile polymer memory with solution-blended ferroelectric polymer/ high-*k* insulators for low voltage operation. *Adv. Funct. Mater.* **2013,** *23* (44), 5484–5493.

Ihlefeld, J.; Laughlin, B.; Hunt-Lowery, A.; Borland, W.; Kingon, A.; Maria, J.-P. Copper compatible barium titanate thin films for embedded passives. *J. Electroceramics* **2005,** *14* (2), 95–102.

Kao, K. C. *Dielectric Phenomena in Solids*. Academic Press: Cambridge, 2004.

Kim, K. J.; Tadokoro, S. *Electroactive polymers for robotic applications, Artificial Muscles and Sensors*. Springer: New York, USA, 2007.

Kim, S. H.; Hong, K.; Xie, W.; Lee, K. H.; Zhang, S.; Lodge, T. P.; Frisbie, C. D. Electrolyte-gated transistors for organic and printed electronics. *Adv. Mater.* **2013,** *25* (13), 1822–1846.

Kittel, C. *Introduction to Solid State Physics*. Wiley: New Jersey, USA, 2005.

Landau, L. D.; Lifshitz, E. M. *Electrodynamics of Continuous Media*. Elsevier: Netherlands, 1984.

Lau, J. H. *Chip on Board: Technology for Multichip Modules.* Springer Science and Business Media; 1994.

Levy, D. H.; Scuderi, A. C.; Irving, L. M. Methods of making thin film transistors comprising zinc-oxide-based semiconductor materials and transistors made thereby, U.S. Patent 7,402,506, 2008.

Li, X.; Niitsoo, O.; Couzis, A. Electrostatically assisted fabrication of silver dielectric core/ shell nanoparticles thin film capacitor with uniform metal nanoparticle distribution and controlled spacing. *J. Colloid Interface Sci.* **2016,** *465*, 333–341.

Lin, Y. S.; Puthenkovilakam, R.; Chang, J. Dielectric property and thermal stability of HfO_2 on silicon. *Appl. Phys. Lett.* **2002,** *81* (11), 2041–2043.

Liu, S.; Xue, S.; Zhang, W.; Zhai, J.; Chen, G. Significantly enhanced dielectric property in pvdf nanocomposites flexible films through a small loading of surface-hydroxylated $Ba_{0.6}Sr_{0.4}TiO_3$ nanotubes. *J. Mater. Chem. A* **2014,** *2* (42), 18040–18046.

Lu, J.; Moon, K.; Xu, J.; Wong, C. Synthesis and dielectric properties of novel high-k polymer composites containing in-situ formed silver nanoparticles for embedded capacitor applications. *J. Mater. Chem.* **2006,** *16*, 1543–1548.

Lu, J.; Moon, K.-S.; Wong, C. Silver/Polymer Nanocomposite as a High-k Polymer Matrix for Dielectric Composites with Improved Dielectric Performance. *J. Mater. Chem.* **2008,** *18*, 4821–4826.

Nan, C.-W. Physics of Inhomogeneous Inorganic Materials. *Progress Mater. Sci.* **1993,** *37* (1), 11–16.

Nan, C.-W.; Shen, Y.; Ma, J. Physical Properties of Composites Near Percolation. *Ann. Rev. Mater. Res.* **2010,** *40*, 131–151.

Nalwa, H. S. *Handbook of Low and High Dielectric Constant Materials and Their Applications*. Academic Press: Cambridge, 1999.

Niitsoo, O.; Couzis, A. Facile Synthesis of Silver Core Silica Shell Composite Nanoparticles. *J. Colloid Interface Sci.* **2011,** *354* (2), 887–890.

Paniagua, S. A.; Kim, Y.; Henry, K.; Kumar, R.; Perry, J. W.; Marder, S. R. Surface-Initiated Polymerization from Barium Titanate Nanoparticles for Hybrid Dielectric Capacitors. *ACS Appl. Mater. Interfaces* **2014,** *6* (5), 3477–3482.

Pecharroman, C.; Moya, J. S. Experimental Evidence of a Giant Capacitance in Insulator Conductor Composites at the Percolation Threshold. *Adv. Mater.* **2000,** *12* (4), 294–297.

Pecharroman, C.; Esteban-Betegon, F.; Bartolome, J.-F., Lopez-Esteban, S.; Moya, J. S. New Percolative Batio$_3$ni Composites with a High and Frequency-Independent Dielectric Constant (R 80000). *Advanced Mater.* **2001,** *13* (20), 1541–1544.

Prakash, B. S.; Varma, K. Dielectric Behavior of Ccto/Epoxy and Al-Ccto/Epoxy Composites. *Composit. Sci. Technol.* **2007,** *67* (11), 2363–2368.

Qi, L.; Lee, B. I.; Chen, S.; Samuels, W. D.; Exarhos, G. J. High-Dielectric-Constant Silverepoxy Composites as Embedded Dielectrics. *Adv. Mater.* **2005,** *17* (14), 1777–1781.

Ramesh, S.; Shutzberg, B. A.; Huang, C.; Gao, J.; Giannelis, E. P.; Dielectric Nanocomposites for Integral Thin Film Capacitors: Materials Design, Fabrication and Integration Issues. *IEEE Trans. Adv. Packag.* **2003,** *26* (1), 17–24.

Rao, Y.; Ogitani, S.; Kohl, P.; Wong, C. Novel Polymer Ceramic Nanocomposite Based on High Dielectric Constant Epoxy Formula for Embedded Capacitor Application. *J. Appl. Polymer Sci.* **2002,** *83* (5), 1084–1090.

Rao, Y.; Wong, C.; Xu, J. Ultra High k Polymer Metal Composite for Embedded Capacitor Application. US Patent 6,864,306, 2005.

Schonhals, A.; Kremer, F. *Analysis of Dielectric Spectra. Broadband Dielectric Spectroscopy*. Springer: Berlin, 2003.

Shen, Y.; Lin, Y.; Nan, C. Interfacial Effect on Dielectric Properties of Polymer Nanocomposites Filled with Core/Shell Structured Particles. *Adv. Funct. Mater.* **2007,** *17* (14), 2405–2410.

Tagarielli, C.; Hildick-Smith, R.; Huber, J. Electromechanical Properties and Electrostriction Response of a Rubbery Polymer for Eap Applications. *Int. J. Solids Struct.* **2012,** *49* (23), 3409–3415.

Tchoul, M. M.; Fillery, S. P.; Koerner, H.; Drummy, L. F.; Oyerokun, F. T.; Mirau, P. A.; Durstock, M. F.; Vaia, R. A. Assemblies of Titanium Dioxide-Polystyrene Hybrid Nanoparticles for Dielectric Applications. *Chem. Mater.* **2010,** *22* (5), 1749–1759.

Tummala, R. Electronic Packaging for High Reliability, Low Cost Electronics. *Springer Sci. Business Media* **1999,** *57.*

Tuncer, E.; Sauers, I.; James, D. R.; Ellis, A. R.; Duckworth, R. C. Nanodielectric System for Cryogenic Applications: Barium Titanate Filled Polyvinyl Alcohol. *IEEE Trans. Dielectrics Elect. Insulat.* **2008,** *15* (1), 236–242.

Wang, Y.; Zhou, X.; Chen, Q.; Chu, B.; Zhang, Q. Recent Development of High Energy Density Polymers for Dielectric Capacitors. *IEEE Trans. Dielectrics Elect. Insul.* **2010,** *17* (4), 1036–1042.

Wang, B.; Liang, G.; Jiao, Y.; Gu, A.; Liu, L.; Yuan, L.; Zhang, W. Two-Layer Materials of Polyethylene and a Carbon Nanotube/Cyanate Ester Composite with High Dielectric Constant and Extremely Low Dielectric Loss. *Carbon* **2013,** *54,* 224–233.

Wan, Y.-J.; Yang, W.-H.; Yu, S.-H.; Sun, R., Wong, C.-P.; Liao, W.-H. Covalent Polymer Functionalization of Graphene for Improved Dielectric Properties and Thermal Stability of Epoxy Composites. *Composit. Sci. Technol.* **2016,** *122,* 27–35.

Wersing, W. High-Frequency Ceramic Dielectrics and Their Application for Microwave Components. *Electronic Ceramics* **1991,** *1991,* 67119.

Wu, H.; Gu, A.; Liang, G.; Yuan, L. Novel Permittivity Gradient Carbon Nanotubes/Cyanate Ester Composites with High Permittivity and Extremely Low Dielectric Loss. *J. Mater. Chem.* **2011,** *21* (38), 14838–14848.

Wu, C.; Huang, X.; Xie, L.; Wu, X.; Yu, J.; Jiang, P. Morphology-Controllable Grapheme Tio_2 Nanorod Hybrid Nanostructures for Polymer Composites with High Dielectric Performance. *J. Mater. Chem.* **2011,** *21* (44), 17729–17736.

Wu, C.; Huang, X.; Wu, X.; Yu, J.; Xie, L.; Jiang, P. TiO_2-Nanorod Decorated Carbon Nanotubes for High-Permittivity and Low-Dielectric-Loss Polystyrene Composites. *Composit. Sci. Technol.* **2012,** *72* (4), 521–527.

Xie, L.; Huang, X.; Wu, C.; Jiang, P. Core-Shell Structured Poly (Methyl Methacrylate)/ $Batio_3$ Nanocomposites Prepared by In Situ Atom Transfer Radical Polymerization: A Route to High Dielectric Constant Materials with the Inherent Low Loss of the Base Polymer. *J. Mater. Chem.* **2011,** *21* (16), 5897–5906.

Xu, H.; Cheng, Z.; Olson, D.; Mai, T.; Zhang, Q.; Kavarnos, G. Ferroelectric and electromechanical properties of poly (vinylidenefluoridetrifluoroethylene chlorotrifluoroethylene) terpolymer. *Appl. Phys. Lett.* **2001,** *78* (16), 2360–2362.

Xu, T.; Cheng, Z.; Zhang, Q. High-Performance Micromachined Unimorph Actuators Based on Electrostrictive Poly(Vinylidene Fluoridetrifluoroethylene) Copolymer. *Appl. Phys. Lett.* **2002,** *80* (6), 1082–1084.

Yang, W.; Yu, S.; Sun, R.; Du, R. Nano-and Microsize Effect of Ccto Fillers on the Dielectric Behavior of Ccto/Pvdf Composites. *Acta Mater.* **2011,** *59* (14), 5593–5602.

Yang, K.; Huang, X.; Huang, Y.; Xie, L.; Jiang, P. Fluoropolymer @ $BaTiO_3$ Hybrid Nanoparticles Prepared Via Raft Polymerization: Toward Ferroelectric Polymer Nanocomposites with High Dielectric Constant and Low Dielectric Loss for Energy Storage Application. *Chem. Mater.* **2013,** *25* (11), 2327–2338.

Yousefi, N.; Sun, X.; Lin, X.; Shen, X.; Jia, J.; Zhang, B.; Tang, B.; Chan, M.; Kim, J.-K. Highly Aligned Graphene/Polymer Nanocomposites with Excellent Dielectric Properties for High performance Electromagnetic Interference Shielding. *Adv. Mater.* **2014,** *26,* 5480–5487.

Yuan, J.-K.; Yao, S.-H.; Dang, Z.-M.; Sylvestre, A.; Genestoux, Bail, J. Giant Dielectric Permittivity Nanocomposites: Realizing True Potential of Pristine Carbon Nanotubes in Polyvinylidene Fluoride Matrix Through an Enhanced Interfacial Interaction. *J. Phys. Chem. C* **2011,** *115* (13), 5515–5521.

Yuan, J.-K.; Yao, S.-H.; Li, W.; Sylvestre, A.; Bai, J. Vertically Aligned Carbon Nanotube Arrays on Sic Microplatelets: A High Figure-Of-Merit Strategy for Achieving Large Dielectric Constant and Low Loss In Polymer Composites. *J. Phys. Chem. C* **2014,** *118* (40), 22975–22983.

Yao, S.-H.; Yuan, J.-K.; Gonon, P.; Bai, j.; Pairis, S.; Sylvestre, A. Effect of Oxygen Vacancy on The Dielectric Relaxation of $Batio_3$ Thin Films in a Quenched State. *J. Appl. Phys.* **2012,** *111* (10), 104–109.

Zaumseil, J.; Sirringhaus, H. Electron and Ambipolar Transport in Organic Field-Effect Transistors. *Chem. Rev.* **2007,** *107* (4), 1296–1323.

Zhang, Q.; Bharti, V.; Zhao, X. Giant Electrostriction and Relaxor Ferroelectric Behavior in Electron-Irradiated Poly (Vinylidene Fluoridetrifluoroethylene) Copolymer. *Science* **1998,** *280* (5372), 21012104.

Zhang, Q.; Li, H.' Poh, M.; Xia, F.; Cheng, Z.; Xu, H.; Huang, C. An All-Organic Composite Actuator Material with a High Dielectric Constant. *Nature* **2002,** *419* (6904), 284–287.

Zhang, S.; Chu, B.; Neese, B.; Ren, K.; Zhou, X.; Zhang, Q. Direct Spectroscopic Evidence of Field-Induced Solid-State Chain Conformation Transformation in a Ferroelectric Relaxor Polymer. *J. Appl. Phys.* **2006,** *99* (4), 044107.

Zhang, S.; Wang, H.; Wang, G.; Jiang, Z. Material with High Dielectric Constant, Low Dielectric Loss, and Good Mechanical and Thermal Properties Produced Using Multi-Wall Carbon Nanotubes Wrapped with Poly (Ether Sulphone) in a Poly (Ether Ether Ketone) Matrix. *Appl. Phys. Lett.* **2012,** *101* (1), 012904.

Zhou, T.; Zha, J.-W.; Hou, Y.; Wang, D.; Zhao, J.; Dang, Z.-M. Surface-Functionalized Mwnts with Emeraldine Base: Preparation and Improving Dielectric Properties of Polymer Nanocomposites. *ACS Appl. Mater. Interfaces* **2011,** *3* (12), 4557–4560.

Zhu, L. Exploring Strategies for High Dielectric Constant and Low Loss Polymer Dielectrics. *J. Phys. Chem. Lett.* **2014,** *5* (21), 36773687.

Zou, C.; Kushner, D.; Zhang, S. Wide Temperature Polyimide/ZrO_2 Nanodielectric Capacitor Film with Excellent Electrical Performance. *Appl. Phys. Lett.* **2011,** *98* (8), 082905.

CHAPTER 4

Selection of High-k Dielectric Materials

N. P. MAITY and RESHMI MAITY

Department of Electronics and Communication Engineering, Mizoram University (A Central University), Aizawl 796004, India

Corresponding author. E-mail: maity_niladri@rediffmail.com

ABSTRACT

It is not straightforward to substitute SiO_2 material with an alternative gate oxide material. The characteristics and properties of the material should be thoroughly studied before use. This chapter describes the properties of promising high-k dielectric materials and their selection for future CMOS technology. The need of high-k material, the required characteristics of the materials, and the materials available satisfying the characteristics are also being analyzed for selection of the high-k dielectric material.

4.1 INTRODUCTION

Finding a substitute gate material with high dielectric constant for upcoming CMOS (complementary metal-oxide-semiconductor) technology generations is one of the greatest challenging complications in the incessant development of microelectronics/nanoelectronics (Iwai and Ohmi, 2002). In order to decrease the gate leakage current, high-k gate dielectric materials are anticipated to substitute SiO_2 in future advanced MOSFET (metal–oxide–semiconductor field-effect-transistor) technology. Acquaint with a physically thicker high-k material can decrease the leakage current to the acceptable limit. Two of the greatest fundamental quantities that are essential to be considered are the dielectric constant and the energy band offset values among the conduction band of the oxide materials and the silicon substrate (Robertson, 2004).

It is not straightforward to substitute SiO_2 with an alternative gate dielectric material. The required properties of gate dielectrics should be systematically considered to provide the key guidelines for selecting an alternative gate dielectric. The dielectric constant of candidate oxides tends to vary inversely with the bandgap, and too high-*k* value cannot be accepted (Robertson, 2002). There are of course oxides with extremely large dielectric values, such as ferroelectrics like $BaTiO_3$, but these have too low bandgap. In fact, a very high dielectric value is unwanted in MOS (metal–oxide–semiconductor) modeling (Schuegraf and Hu, 1994). In view of the fact that the significant motivation for replacing SiO_2 material with high-*k* materials is the leakage current reduction; accurate analytical modeling of the leakage current is essential to understand the tunneling phenomenon and MOS devices scaling limits to make guaranteed that the selected high-*k* dielectric materials are very much scalable and functional for numerous future generations of CMOS technology.

4.2 NEED OF HIGH-k DIELECTRIC MATERIALS

High-*k* dielectric materials are a technology breakthrough as elementary as to encourage important questions regarding the forthcoming financial side of the manufacturing industry. The aforementioned is not the objective of a material transform in the gate dielectric. High-*k* dielectric materials are also coming up with modifications to the physical structure of the MOSFET technology itself as well as the substrate. The beginning of the planar process in the late 1950s has the semiconductor industry encountered such a fundamental transformation in very large-scale integration (VLSI) fabrication technology. In all probability, these modifications will perhaps mark the ending of the conventional planar, poly process, consequently the requirement for an assessment of Moore's law (Huff and Gilmer, 2005), which has forever been solved with novel semiconductor technology. It provides an establishment for future economic development and as such, sets the stage for a technical dissertation on high-*k* dielectric materials.

The accomplishment of high-*k* gate dielectrics is one of quite a lot of strategies developed to permit further miniaturization of microelectronic/

nanoelectronic components. Replacing the recent SiO_2 gate dielectric with a high-*k* dielectric material permits increased gate capacitance value without the associated leakage current effects. The direct tunneling current with conventional SiO_2 material can be significantly suppressed by using thicker high-*k* insulator of MOS devices with the same effective oxide thickness as shown in Figure 4.1, that is, to avoid direct tunneling phenomenon, a material with a higher dielectric constant value can be made thick sufficient and still continue to maintain the capacitance of the thinner SiO_2 film (Michael, 2003). Figure 4.2 demonstrates the comparison between conventional SiO_2 gate dielectric material structure and a high-*k* dielectric material structure (Moore, 2003).

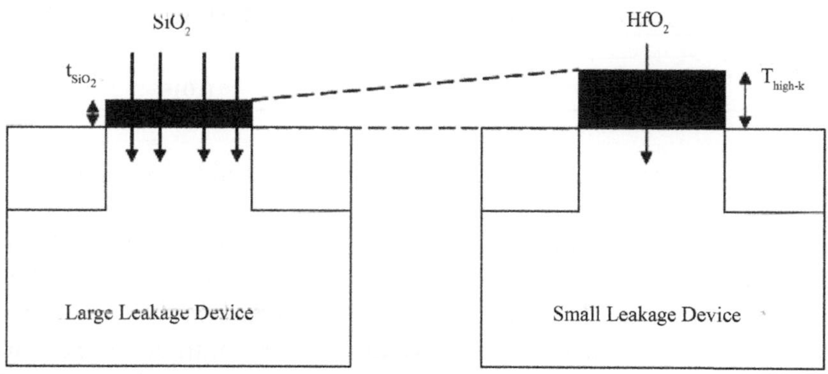

FIGURE 4.1 Suppression of the direct tunnel leakage current through the gate insulator by introducing alternative gate oxides having a high dielectric constant.

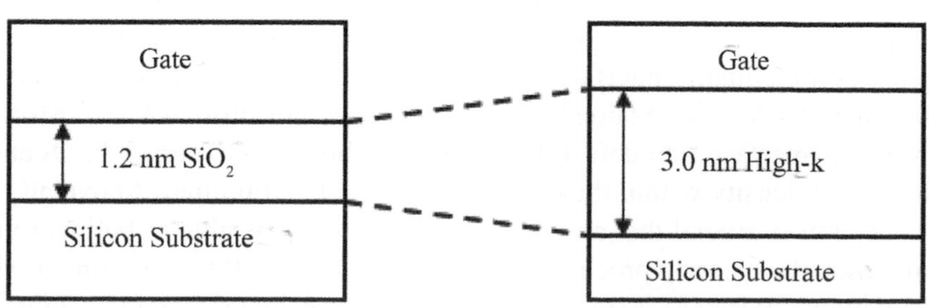

FIGURE 4.2 Comparison between conventional SiO_2 gate dielectric structure and a high-*k* dielectric structure.

The comparison between SiO_2 gate dielectric with high-*k* dielectric material in terms of the leakage current and capacitance value is also shown in Table 4.1 (Moore, 2003). It is nothing but the scaling pressures that have tainted the favored gate dielectric material from the established SiO_2 to high-*k* materials in an attempt to extend the lifetime of the presently used technologies. In view of the fact that SiO_2 approached its physical limit, so there are a number of problems arising, this does not permit further scaling. Hence, to alleviate the problems, alternative high-*k* dielectric materials have to be introduced to meet the rigorous requirements.

TABLE 4.1 Comparison between SiO_2 Gate Dielectric with High-*k* Dielectric Materials in Terms of the Leakage Current and Capacitance Value.

Parameters	SiO_2	High-*k*
Capacitance	1[a]	~1.6[a]
Leakage	1[a]	~< 0.01[a]

The high-*k* material should have a higher permittivity than that of SiO_2, and bandgap must be large enough so that there is large band offsets to give low leakage currents. It is very much important that the high-*k* dielectric material should be stable when in contact with Si. Most of these materials were found to form an unwanted interfacial layer by reacting with Si. Hence, it reduces the effective dielectric permittivity of the structure. Thus, it is to be noted while choosing a material that it is more stable when in contact with the Si or the other way is to deposit a barrier layer on Si and then deposit actual dielectric on top of it. Although the majority of the metal oxides of choice are greatly crystalline in nature, materials having amorphous structure are attractive. The crystalline dielectrics are problematic for the reason that the grain boundaries serve as high leakage paths. A high quality interface is always desirable since it would preserve the capacitance gain obtained by using a high-*k* dielectric material. Moreover, the interface trap defect density should be equal to current SiO_2 and the defect density within the oxide layer should be minimal. Above all, it is very much crucial that any new material should be able to be incorporated into the existing process technology (Michael, 2003; Persson, 2004; Gaddipati, 2004). As a result, a gate dielectric material with a higher dielectric constant affords a greater physical thickness and achieves the equivalent capacitance as achieved by SiO_2 with lesser thickness. In other words by using high-*k* gate dielectric material, we can employ a thicker gate layer with lesser equivalent oxide thickness.

4.3 MATERIAL REQUIREMENTS FOR HIGH-k GATE DIELECTRIC

Till date, on the other hand, there is no particular material that is talented to satisfy all necessities for an ultimate gate oxide. It is critical to use a material with a high-*k* value to substitute SiO_2 as the gate dielectric material. Major requirements comprise (Wilk et al., 2001; Houssa, 2004; Hori, 1997) of:

- high dielectric constant and large bandgap;
- high band offset with electrodes;
- thermally and chemically stable in contact with semiconductor substrate;
- scalable equivalent oxide thickness EOT < 15 Å;
- compatibility with gate electrode material;
- density of interface states comparable to SiO_2;
- low lattice mismatch and similar thermal expansion coefficient with Si;
- mobility comparable to SiO_2;
- negligible capacitance voltage hysteresis (<20 mV);
- good reliability (no charge trapping, high breakdown voltage etc.);
- long lifetime;
- good kinetic stability;
- low high-*k*/Si interface state density ($\sim 10^{10}\ cm^{-2}eV^{-1}$);
- low fixed charge density ($\sim 10^{10}\ cm^{-2}eV^{-1}$); and
- high enough channel carrier mobility (~90% of Si–SiO_2 system).

4.4 ALTERNATIVE HIGH-k DIELECTRIC MATERIALS

There are several alternative high-*k* dielectric materials studied for the determination of replacing SiO_2 in future technology due to the problems of high gate leakage current, standby power consumption, and gate oxide reliability. Since similar capacitance with a physically larger thickness can be provided by the high-*k* films, therefore they can be used to reduce the gate leakage current (Jiang et al., 2004). Some of them are aluminum oxide (Al_2O_3), titanium oxide (TiO_2), titanium-based compound (TiLaO), tantalum oxide (Ta_2O_5), silicon nitride (Si_3N_4), hafnium oxide (HfO_2), hafnium silicate ($HfSiO_4$), hafnium-based compounds (HfON, HfLaO, HfTaO, HfTiSiO, HfAlO, HfSiO, HfSiON, and HfAlON), zirconium dioxide (ZrO_2), zirconium silicate ($ZrSiO_4$), zirconium-based compound (ZrON), lead titanate ($PbTiO_3$), cerium oxide (CeO_2), dysprosium oxide (Dy_2O_3), strontium titanium oxide (Sr_2TiO_4), strontium zirconate ($SrZrO_3$), strontium titanate [$SrTiO_3$ (STO)], strontium-based compounds ($SrTa_2O_6$ and $SrHfO_3$), barium strontium titanate [$Ba_xSr_{1-x}TiO_3$(BST)], titanium–aluminum oxynitride (TAON), titanium–aluminum oxide (TAO), yttrium oxide (Y_2O_3), lanthanum oxide (La_2O_3), lanthanum aluminate ($LaAlO_3$), lanthanum lutetium oxide ($LaLuO_3$), lutetium oxide (Lu_2O_3), erbium oxide (Er_2O_3), gadolinium oxide (Gd_2O_3), praseodymium oxide (Pr_2O_3) etc. (Gaddipati, 2004; Jiang, 2004; Miyoshi et al., 2010; Demkov and Navrotsky, 2005; Yeh et al., 2009; Liu et al., 2011; Noor et al., 2010; Cheng et al., 2011; Maity et al., 2016; 2017a; 2017b; Zhang et al., 2011).

Si_3N_4 and Al_2O_3 have comparatively low-*k* (low dielectric constant) value compared to the other proposed materials. Al_2O_3 is one of the first systems that have been studied to substitute SiO_2 as gate dielectric. Even though, Al_2O_3 has a lower permittivity, its larger bandgap makes it a good quality barrier layer in MOS structures. Recently, several high-*k* dielectric materials have been considered as shown in Table 4.2 and show outstanding result. TiO_2 and $BaSrTiO_3$ have their dielectric constant values highest among all dielectric materials. Conversely, thermodynamic studies point to that Ta and Ti materials are thermodynamically unstable in contact with silicon at characteristic device processing temperatures (~1000°C). Ta_2O_5 was considered a talented high-*k* contender, but then again the low barrier (0.3 eV) indicates considerable gate leakage current owing to electron emission into the conduction band, which is not desirable and acceptable for future technology. But it has been effectively implemented in dynamic random access memory (DRAM) designs (Jiang, 2004; Brezeanu et al., 2010; Westlinder, 2004).

Binary and ternary oxides (HfAlO, HfSiO, HfSiON, and ZrON) have, in general, better crystallization temperature compared to single metal oxides such as Al_2O_3, Ta_2O_5, ZrO_2, and HfO_2. But these frequently demonstrate unwanted effect such as phase separation at high temperatures. By incorporating nitrogen, these oxides can obtain over phase separation and continue in amorphous state at high temperature, even as keeping barrier height sufficiently high. So far, HfSiON and HfO_2 with appropriate treatments have been successfully established to the choice of modern MOS technology. Promising alternative high-*k* materials is the multi-component gate dielectrics based on a diversity of metal oxides (Angelov et al., 2011).

TABLE 4.2 Conduction Band Offset and Valence Band Offset of Some High-*k* Dielectric Materials.

Materials	Conduction band offset to Si (eV)	Valence band offset (eV)
SiO_2	3.5	4.4
Si_3N_4	2.4	1.8
Al_2O_3	2.8	4.9
$HfSiO_4$	1.5	3.6
Y_2O_3	2.3	2.6
$ZrSiO_4$	1.5	3.4
HfO_2	1.5	3.3
ZrO_2	1.4	3.2
Ta_2O_5	0.3	2.95
La_2O_3	2.3	2.6
TiO_2	0.0	2.4

Among candidates, gate metal oxides, such as ZrO_2 and HfO_2, and Hf-based silicates are reasonably promising in the meantime as they have superior dielectric properties and respectable quality of thermal stability on silicon substrates. Morcover, amorphous ZrO_2 has been developed and outstanding properties are observed. Both materials ZrO_2 and HfO_2 have a dielectric constant of about 22 (Silvaco Inc., 2015; Maity et al., 2017; 2016; 2018) and a large bandgap. HfO_2 is considered as the most promising, or more accurately, most explored high-*k* dielectric material of nowadays due to its low tunneling current, high enough conduction bands offset, and satisfactory thermal stability. Barium strontium titanate (BST) shows overwhelmingly higher permittivity, and was reported not to be thermally stable with Si substrates. It appears to be suitable because the very high dielectric constant causes field-induced barrier lowering (FIBL) that degrades short channel effects of MOS transistor.

Some other high-*k* materials including lithium oxide (Li_2O), beryllium oxide (BeO), magnesium oxide (MgO), calcium oxide (CaO), strontium oxide (SrO), scandium oxide (Sc_2O_3), thorium oxide (ThO_2), uranium oxide (UO_2), neodymium oxide (Nd_2O_3), barium zirconate ($BaZrO_3$), ruthenium oxide (RuO_2), and strontium ruthenium oxide ($SrRuO_3$) also may provide good outcome (Capodieci, 2005; Kumar and Subba Rao, 2011).

4.5 SELECTED HIGH-k DIELECTRIC MATERIALS

A number of high-*k* materials have been considered and show excellent result. Nevertheless, they are not the complete solution. Although high-*k* dielectric materials appearance is very promising, but there are certain challenges and concerns that have to be encountered before successful transition from SiO_2 to high-*k* materials. Novel materials with high dielectric constant are needed to keep away from unexpected and expensive technological and manufacturing process changes, decrease leakage, and get better performance. In this regard, Al_2O_3, ZrO_2, and HfO_2 (Robertson and Wallace, 2015; Maity et al., 2016) have been selected for the analysis.

4.5.1 Aluminum Oxide

- The higher dielectric constant of Al_2O_3 ($k \sim 9.3$) compared to SiO_2 allows device operation at a higher electric field, taking advantage of the high breakdown field (Tanner et al., 2007; Maity et al., 2016; 2017).
- The large bandgap of Al_2O_3 thin films (E_g = 8.6 eV) relative to other high-*k* oxides potentially enables adequate barrier heights at the interface (Maity et al., 2016; 2017).
- To accomplish high capacitance while keeping the leakage current low, Al_2O_3 is suitable with comparatively thick films. These thicker films must prevent the excessive tunneling currents detected for very thin SiO_2 films (Groner and George, 2003).
- This material can allow us to reduce the size of MOS devices, because it would be able to withstand higher voltages for the same gate size. The thermodynamic stability of Al_2O_3 is suitable, has high bandgap energy, low leakage current, and adequate value of dielectric constant (Groner and George, 2003).

- Al_2O_3 gate dielectrics displayed nearly same degradation rate of capacitance reduction as SiO_2 with increment of frequency, and preserved higher *k* quality better than SiO_2 over a very wide range of frequency (Chen et al., 2002; Maity et al., 2016; 2017).

4.5.2 Zirconium Oxide

- Most of the high-*k* materials are not thermally stable directly on Si, due to the formation of metal silicides in the course of fabrication. ZrO_2 is one of the most attractive candidates due to its high-*k* values, wide bandgap, and good thermal stability in contact with Si due to their excellent properties (Jeong et al., 2005; Maity et al., 2014; 2017a; 2017b).
- ZrO_2 is a polymorphic material and occurs in three forms: monoclinic, tetragonal, and cubic. The monoclinic phase is stable at room temperatures up to 1170°C; the tetragonal phase is stable at temperatures in the range of 1170–2370°C, and the cubic phase is stable at over 2370°C (Chevalier et al., 2009).
- Thermally stable and has stable electrical interface with silicon surface (Gupta et al., 2012).
- The value of dielectric constant is high, so it can be used for long time of year of scaling (Gupta et al., 2012).
- It is compatible with processing technology (Gupta et al., 2012; Maity et al., 2014; 2017).

4.5.3 Hafnium Oxide

- HfO_2 is a promising candidate due to its thermodynamic stability on Si, high dielectric constant, and relatively large bandgap (Wang, 2010; Maity et al., 2016; 2017; 2014; 2019a; 2019b; 2018).
- Some materials with higher heat of formation than SiO_2 may also be slightly reactive with Si such as ZrO_2, forming the silicide, $ZrSi_2$. Among these high-*k* dielectrics, HfO_2 has both a high-*k* value as well as chemical stability with water and Si (Gupta et al., 2012).
- HfO_2 is the most promising dielectric after SiO_2 due to its high dielectric constant and very high breakdown voltage. A good

quality and desired thickness can be easily achieved in laboratory that is why HfO_2-based MOS device seems to be as future devices (Gupta et al., 2012; Maity et al., 2014; 2016; 2017).

- Pure forms of HfO_2 and its silicates are the potential candidates. A continuous research and development on this material considerably seems to be more advanced compared to other high-*k* dielectrics (Gupta et al., 2012; Maity et al., 2014; 2016; 2017).
- HfO_2 has received significant attention in recent years as a potential replacement for SiO_2 as the gate dielectric material in complementary metal oxide semiconductor CMOS technology due to its high dielectric constant (Wang, 2010).
- The use of a gate dielectric with an increased *k* allows a reduction in driving voltage of a transistor or an increase in dielectric film thickness while maintaining the same gate capacitance, thus suppressing the gate leakage current due to electron tunneling (Wang, 2010).
- An amorphous gate dielectric material is preferable for better device stability, reduced leakage currents, and improved device-to-device uniformity. However, it has been suggested that amorphous HfO_x is similar in local atomic coordination to the monoclinic polymorph and has a limited *k* of only ~22 (Wang, 2010; Maity et al., 2014; 2016; 2017).
- The choice of alternative gate dielectrics has been narrowed to HfO_2, ZrO_2, and their silicates due to their excellent electrical properties and high thermal stability in contact with Si (Huang et al., 2010; Maity et al., 2014; 2016; 2017a; 2017b).

4.6 SUMMARY

In this chapter, the need of high-*k* materials, the required characteristics of the materials, and the materials available satisfying the characteristics are being analyzed for the selection of high-*k* dielectric materials.

KEYWORDS

- **high-*k* materials**
- **hafnium oxide**
- **zirconium oxide**
- **aluminum oxide**
- **material properties**

REFERENCES

Angelov, G.; Bonev, N.; Rusev, R.; Hristov, M.; Paskaleva, A.; Spassov, D. Verilog-A Model of a High-k HfO_2-Ta_2O_5 Capacitor. In *Proc. Int. Conf. of Mixed Design of Integrated Circuits and Systems*, 2011.

Brezeanu, G.; Brezeanu, M.; Bernea, F. High-K Dielectrics in Nano and Microelectronics, [Online]. www.romnet.net/ro/seminar16martie2010/, 2010.

Capodieci, V. Molecular Beam Deposition (MBD) and Characterization of High-k Material as Alternative Gate Oxides for MOS-Technology. Ph.D. Dissertation, Faculty of Electronics and Information Technology, Bundeswehr Munchen University, 2005.

Cheng, C.; Horng, J.; Tsai, H. Electrical and physical characteristics of HfLaON-gated metal–oxide-semiconductor capacitors with various nitrogen concentration profiles. *Microelectron. Eng.* **2011,** *88*, 159–165.

Chen, S.; Lai, C.; Chan, K.; Chin, A.; Hsieh, J.; Liu, J. Frequency-dependent capacitance reduction in high-k $AlTiO_X$ and Al_2O_3 gate dielectrics from IF to RF frequency range. *IEEE Electron Dev. Lett.* **2002,** *23*, 203–205.

Chevalier, J.; Gremillard, L.; Virkar, A., Clarke, D. The tetragonal-monoclinic transformation in zirconia: Lessons learned and future trends. *J. Am. Ceramic Soc.* **2009,** *92,* 1901–1920.

Demkov, A. A.; Navrotsky, A. *Materials Fundamentals of Gate Dielectrics.* Springer: Netherlands, 2005.

Gaddipati, S. Characterization of HfO_2 Films for Flash Memory Applications. Ph.D. Dissertation, University of South Florida, USA, 2004.

Groner, M. D.; George, S. M. High-*k* dielectrics grown by atomic layer deposition: capacitor and gate applications. in *Interlayer Dielectrics for Semiconductor Technologies*; 1st ed., Muraka, S., Eizenberg, M. Sinha, A. K. Eds.; Academic Press (Elsevier): USA, 2003; pp 327–348.

Gupta, S. K.; Singh, J.; Akhtar, J. Material Processing for Gate Dielectrics on Silicon Carbide (SiC) Surface. In *Physics and Technology of Silicon Carbide Devices*; Hijika, Y., Ed.; InTech: Croatia, 2012; pp 207–234.

Hori, T. *Gate Dielectrics and MOS ULSIs: Principles, Technologies, and Applications.* Springer: Germany, 1997.

Houssa, M. *High-k Gate Dielectrics.* Institute of Physics Publishing: Bristol, UK, 2004.

Huang, A. P.; Yang, Z. C.; Chu, P. K. Hafnium-based High-k Gate Dielectrics. in *Advances in Solid State Circuits Technologies*; Chu, P. K., Ed.; InTech: Croatia, 2010; pp 333–350.010, 333–350.

Huff, H. R.; Gilmer, D. C. *High Dielectric Constant Materials VLSI MOSFET Applications.* Springer: New York, 2005.

Iwai, H.; Ohmi, S. Silicon integrated circuit technology from past to future. *Microelectron. Reliab.* **2002,** *42*, 465–491.

Jiang, J. Studies of Thin Silicon Oxides and High-k Materials for Gate Dielectrics in Metal-Insulator-Semiconductor Structures. Ph.D. Dissertation, The Pennsylvania State University, USA, 2004.

Jeong, S.; Bae, I.; Shin, Y.; Lee, S.; Kwak, H.; Boo, J. Physical and Electrical Properties of ZrO_2 and YSZ High-k gate dielectric thin films grown by RF magnetron sputtering. *Thin Solid Films* **2005,** *475*, 354–358.

Kumar, B. R.; Subba Rao, T. High-k Gate Dielectrics of Thin Films with its Technological Applications – A Review. *Int. J. Pure Appl. Sci. Technol.* **2011,** *4*, 105–114.

Liu, Y.; Shen, S.; Brillson, L.; Gordon, R. Impact of ultrathin Al_2O_3 barrier layer on electrical properties of $LaLuO_3$ metal-oxide-semiconductor devices. *Appl. Phys. Lett.* **2011,** *98*, 122907.

Maity, N. P.; Maity, R.; Thapa, R. K.; Baishya, S. Study of Interface Charge Densities for ZrO_2 and HfO_2 Based Metal Oxide Semiconductor Devices. *Adv. Mater. Sci. Eng.* **2014,** *2014*, 1–6.

Maity, N. P.; Maity, R.; Thapa, R. K.; Baishya, S.A Tunneling Current Density model for Ultra Thin HfO_2 High-k Dielectric Material Based MOS Devices. *Superlattice. Microstruct.* **2016,** *95*, 24–32.

Maity, N. P.; Thakkur, R. R.; Maity, R.; Thapa, R. K.; Baishya, S. Analysis of Interface Charge Densities for High-k Dielectric Materials based Metal-Oxide-Semiconductor Devices. *Int. J. Nanosci.* **2016,** *15*, 1660011 1-6.

Maity, N. P.; Maity, R.; Baishya, S. Voltage and Oxide Thickness Dependent Tunneling Current Density and Tunnel Resistivity Model: Application to High-k Material HfO_2 Based MOS Devices. *Superlattice. Microstruct.* **2017,** *111*, 628–641.

Maity, N. P.; Maity, R.; Thapa, R. K.; Baishya, S. Influence of Image force effect on tunnelling current density for high-k material ZrO_2 ultra thin films based MOS Devices. *J. Nanoelectron. Optoelectron.* **2017,** *12* (1), 67–71.

Maity, N. P.; Maity, R.; Baishya, S. A. Tunneling Current Model with a Realistic Barrier for Ultra Thin High-k Dielectric ZrO_2 Material based MOS Devices. *Silicon* **2018,** 10 (4), 1645–1652.

Maity, N. P.; Maity, R.; Baishya, S. An analytical model for the surface potential and threshold voltage of a double-gate heterojunction tunnel FinFET. *J. Comput. Electron.* [Online early access]. DOI: 10.1007/s10825-018-01279-5. Published Online: 2018.

Maity, N. P.; Maity, R.; Maity, S.; Baishya, S. Comparative analysis of the Quantum FinFET and Trigate FinFET based on Modeling and Simulation. *J Comput. Electron.* [Online early access]. DOI: 10.1007/s10825-018-01294-z. Published Online: 2019

Jones, M. N. Leakage Current Behavior of Reactive RF Sputtered HfO_2 Thin Films. M.S. Dissertation, University of Florida, USA, 2003.

Miyoshi, J.; Diniz, J.; Barros, A.; Doi, I.; Von Zuben, A. Titanium–aluminum oxynitride (TAON) as high-k gate dielectric for sub-32 nm CMOS technology. *Microelectron. Eng.* **2010,** *87*, 267–270.

Moore, G. E. No Exponential is Forever: But Forever can be Delayed. in *Proc. IEEE International Solid-State Circuits Conference*, 2003.

Noor, F. A; Abdullah, M.; Khairurrijal, S.; Ohta, A.; Miyazaki, S. Electron and hole components of tunneling currents through an interfacial oxide-high-*k* gate stack in metal-oxide-semiconductor capacitors. *J. Appl. Phys.* **2010,** *108*, 093711.

Persson, S. Modeling and Characterization of Novel MOS devices. Ph.D. Dissertation, Solid State Device Laboratory, Royal Institute of Technology, Sweden, 2004.

Robertson, J. Band Offsets of Wide-Band-Gap Oxides and Implications For Future Electronic Devices. *J. Vacuum Sci. Technol.* **2000,** *B 18*, 1785.

Robertson, J. High Dielectric Constant Oxides. *Eur. Phys. J. Appl. Phys.* **2004,** 28, 265–291.

Robertson, J.; Wallace, R. High-k Materials and Metal gates for CMOS applications. *Mater. Sci. Eng. R* **2015,** *88*, 1–41.

Schuegraf, K. F.; Hu, C. Hole Injection SiO_2 Breakdown model for very low voltage lifetime extrapolation. *IEEE Trans. Electron Dev.* **1994,** *41*, 761–767.

Silvaco Inc. *ATLAS user's Manual*. Silvaco Inc.: Santa Clara, USA, 2013.

Tanner, C. M.; Perng, Y.; Frewin, C.; Saddow, S.; Chang, J. P. Electrical performance of Al_2O_3 Gate Dielectric Films Deposited by Atomic Layer Deposition on 4H-SiC. *Appl. Phys. Lett.* **2007,** *91*, 203510-203510-3.

Wang, X. Simulation Study of Scaling Design, Performance Characterization, Statistical Variability and Reliability of decananometer MOSFETs. Ph.D. Dissertation, Department of Electronics and Electrical Engineering, University of Glasgow, Glasgow, UK, 2010.

Westlinder, J. Investigation of Novel Metal Gate and High-k Dielectric Materials for CMOS Technologies. Ph.D. Thesis, Faculty of Science and Technology, Uppsala University, Sweden, 2004.

Wilk, G. D.; Wallace, R. M; Anthony, J. M. High-k Dielectrics: Current Status and material properties Considerations. *J. Appl. Phys.* **2001,** *89*, 5243–5275.

Yeh, L.; Chang, I.; Chen, C.; Lee, J. Reliability Properties of Metal-Oxide-Semiconductor Capacitors Using $LaAlO_3$ high-k dielectric. *Appl. Phys. Lett.* **2009,** *95*, 162902.

Zhang, P.; Nagle, R.; Deepak, N.; Povey, I.; Gomeniuk, Y.; O'Connor, E.; Petkov, N.; Schmidt, M.; O'Regan, T.; Cherkaoui, K.; Pemble, M.; Hurley, P.; Whatmore, R. The structural and electrical properties of the $SrTa_2O_6/In_{0.53}Ga_{0.47}As$ /InP system. *Microelectron. Eng.* **2011,** *88*, 1054–1057.

CHAPTER 5

Tunneling Current Density and Tunnel Resistivity: Application to High-k Material HfO_2

N. P. MAITY* and RESHMI MAITY

Department of Electronics and Communication Engineering, Mizoram University (A Central University), Aizawl-796004, India

Corresponding author. E-mail: maity_niladri@rediffmail.com

ABSTRACT

This chapter proposes a tunneling current model based on Tsu-Esaki concept where potential energy profile for each region of the gate-dielectric-semiconductor structure is considered. The developed model understands the behavior of the gate current as a function of applied potential, dielectric thickness, and temperature. The Schrödinger's wave equation has been used for the different potential energy regions for estimation of transmission coefficients and tunnel resistivity. Effect of high-k material on the model is also explained.

5.1 INTRODUCTION

For several years the semiconductor technology has gone through drastic change in performance as well as productivity owing to scaling. The metal-oxide-semiconductor (MOS) transistor scaling has gone even more rapidly than the famous Moore's law. This development has already led to length scales where the electrical characteristics of the device are controlled by quantum-mechanical effects. One of the supreme exciting of these belongings is the quantum-mechanical tunneling of charge carriers through classically forbidden regions. The conception of tunneling shows an important factor in the design and improvement of nanoelectronic devices. It is so essential for explanation of tunneling current effects in the modeling and design of semiconductor devices. The tunneling current density in MOS devices is determined by two key magnitudes: the supply function and the transmission coefficient. The supply function explains the supply of existing electrons and it is evaluated using the energy distribution of the electrons. The transmission coefficient defines the probability that an electron can tunnel through the potential barrier. In equilibrium condition the energy distribution can be estimated by a Maxwellian distribution (Das, 2011). As the electric field in real MOS devices is very high, models without Maxwellian have to be considered to explain properly the exact shape of the distribution function. The transmission coefficient can be found by solving the Schrödinger's equation in the region considered and Wentzel–Kramers–Brillouin (WKB) approximation is to be considered (Maity et al., 2016; 2017). But the WKB approximation does not replicate transmission coefficient oscillations that are detected in ultrathin gate dielectrics. For better explanation of tunneling phenomenon through dielectric layer, it is essential to determine the properties of wave function interference that can be accomplished by the transfer-matrix method. Nevertheless, the particular method is inclined to numerical uncertainties. A new encouraging methodology is the quantum transmitting boundary method. It permits a steady and consistent estimation of the transmission coefficient. This consists of the supply function modeling, evaluation of the transmission coefficient, and investigation of tunneling current density.

In solid state technology, SiO_2 is a key material due to its great quality as an insulator material for MOS transistor. SiO_2 material has been used as gate dielectric for decades. It has respectable interface properties and fabrication advantages with silicon. Nonetheless, SiO_2 layer ~2 nm produce severe leakage current dominated by tunneling current owing to direct tunneling of electrons, which affect the device reliability (Maity et al., 2015). The resolution of this problem is to use high dielectric constant material as oxide material to achieve same equivalent capacitance in the continuous improvement of MOS technology. Gate tunneling current decrease is the significant motivation for the replacement of SiO_2 with the high-k materials. These oxide materials must have a lot of prerequisite significant properties, such as good interface properties with silicon and other semiconductors, greater thermodynamic stability, and inferior defect density. These materials should have better breakdown strength at high electric fields, lower diffusion coefficient and obtain high enough band offset values. There are several materials that have been investigated, such as HfO_2, $HfSiO_4$, ZrO_2, $ZrSiO_4$, Al_2O_3, TiO_2, Si_3N_4, CeO_2, $LaAlO_3$, La_2O_3, and Y_2O_3 (Maity et al., 2016; 2017a; 2017b; 2014; 2018; 2019). Then again only very limited number of them have the essential potentials. Among the all high-k material HfO_2 attracted an enormous consideration for modern MOS applications. Amorphous HfO_2 is considered to be one of the promising materials that can replace SiO_2 for forthcoming MOS transistor. HfO_2 has a significant position in recent MOS technology due to its several advantages. Among the materials, gate metal oxide such as HfO_2 is relatively promising having an appropriate dielectric constant of about 22 (Maity et al., 2016; Ganapathi et al., 2016; Fu et al., 2014; Sochacki et al., 2016; Litta et al., 2015; Lin et al., 2006) and a large band gap (~ 5.6 eV) (Yu et al., 2004).

In this chapter, we investigate the WKB approximation and Tsu-Esaki model-based gate current model for ultrathin oxide film-based MOS devices. The transmission coefficient has been determined self-consistently by Schrödinger equations. The potential wave functions have been calculated for different regions at different gate voltages. This is the key to modeling the tunneling current density. The temperature-dependent investigation of the tunneling currents and tunneling resistivity is done for MOS devices with SiO_2 and HfO_2 materials.

5.2 TUNNELING CURRENT DENSITY

The methods of electron tunneling from the electron conduction band (ECB) and holes tunneling from the electron valence band (EVB) can be investigated considering an energy barrier as described in Taur et al. (1997). The semiconductor and metal regions are separated by an energy barrier with barrier height $q\phi_B$ measured from the Fermi energy to the conduction band edge of the insulating layer. The distribution functions at both sides of the barrier are specified (Taur et al., 1997). The dissimilar masses corresponding to the band structure of the considered material are lumped into a single value for the effective mass. This is represented by m_{eff} in the metal and semiconductor and m_{ox} in the oxide layer. The dispersion relation for parabolic bands in semiconductors is approximated by

$$E = \left(\frac{\hbar^2 k^2}{2m_{eff}}\right) = \left[\frac{\hbar^2\left(k_x^{\ 2} + k_y^{\ 2} + k_z^{\ 2}\right)}{2m_{eff}}\right] \tag{5.1}$$

where, ϕ_B is the built in potential, q is electronic charge, $\hbar$ is Plank's constant, $k = (k_x e_x + k_y e_y + k_z e_z)$ (wave vector), e_x, e_y, and e_z are the unit wave vectors. Here only transitions in the x-direction are considered. So the parallel wave vector $k_\rho = \left(k_y e_y + k_z e_z\right)$ is not altered by the tunneling process. Therefore, net tunneling current density from metal to semiconductor can be written as the net difference between current flowing from metal to semiconductor and vice versa (Momose et al., 1996; Lo et al., 1997).

$$J_T = J_{T(1\to 2)} - J_{T(2\to 1)} \tag{5.2}$$

The tunneling current density through the two interfaces depends on the perpendicular component of the wave vector k_x, the transmission coefficient T_c, the perpendicular velocity v_x, the density of states $g_1(k_x)$ and $g_2(k_x)$, the distribution function at both sides of the barrier $f_1(E)$ and $f_2(E)$:

$$dJ_{T(1\to 2)} = \left[qT_C\left(k_x\right)v_x g_1\left(k_x\right)f_1\left(E\right)\right]\left[1 - f_2\left(E\right)\right]dk_x \tag{5.3}$$

$$dJ_{T(2\to 1)} = \left[qT_C\left(k_x\right)v_x g_2\left(k_x\right)f_2\left(E\right)\right]\left[1 - f_1\left(E\right)\right]dk_x \tag{5.4}$$

In these expressions, it is presumed that the transmission coefficient only be determined by the momentum perpendicular to the interface (Taur et al., 1997). The density of k_x states $g(k_x)$ is well-defined by

$$g\left(k_x\right) = \int_0^\infty\int_0^\infty\left[g\left(k_x, k_y, k_z\right)\right]dk_y dk_z \tag{5.5}$$

where, $g(k_x, k_y, k_z)$ indicates the three-dimensional density of states in the momentum space. Let us consider the quantized wave vector components within a cube of side length, L.

$$\Delta k_x = \frac{2\pi}{L}, \qquad \Delta k_y = \frac{2\pi}{L}, \qquad \Delta k_z = \frac{2\pi}{L} \tag{5.6}$$

produces for the density of states within the cube,

$$g\left(k_x,k_y,k_z\right)=2\times\left(\frac{1}{\Delta k_x \Delta k_y \Delta k_z}\right)\times\left(\frac{1}{L^3}\right)=\frac{1}{4\pi^3} \tag{5.7}$$

On behalf of the parabolic dispersion eq (5.2) the velocity and energy components in tunneling direction follow. Therefore,

$$v_x=\left(\frac{1}{\hbar}\times\frac{\partial E}{\partial k_x}\right)=\left(\frac{\hbar^2 k_x^2}{m_{eff}}\right),\qquad E_x=\left(\frac{1}{2}\times\frac{\hbar^2 k_x^2}{m_{eff}}\right),\qquad v_x dk_x=\left(\frac{1}{\hbar}\right)dE_x \tag{5.8}$$

Hence eqs (5.3) and (5.4) become

$$dJ_{T(1\to2)}=\left(\frac{q}{4\pi^3\hbar}\right)T_C(E_x)dE_x\int_0^\infty\int_0^\infty f_1(E)\left[1-f_2(E)\right]dk_y dk_z \tag{5.9}$$

$$dJ_{T(2\to1)}=\left(\frac{q}{4\pi^3\hbar}\right)T_C(E_x)dE_x\int_0^\infty\int_0^\infty f_2(E)\left[1-f_1(E)\right]dk_y dk_z \tag{5.10}$$

Polar coordinates are used for the parallel wave vector components, $k_\rho=\sqrt{k_y^2+k_z^2}$. Here $k_y=k_\rho\cos(\gamma)$, $k_z=k_\rho\sin(\gamma)$ and γ is the angle between the energy in y and z directions and denoted as, $\gamma=\tan^{-1}\left(k_z/k_y\right)$. So the current density evaluates to

$$J_{T(1\to2)}=\left(\frac{4\pi m_{eff}q}{h^3}\right)\int_{E_{min}}^{E_{max}}T_C(E_x)d(E_x)\int_0^\infty f_1(E)\left[1-f_2(E)\right]dE_\rho \tag{5.11}$$

$$J_{T(2\to1)}=\left(\frac{4\pi m_{eff}q}{h^3}\right)\int_{E_{min}}^{E_{max}}T_C(E_x)d(E_x)\int_0^\infty f_2(E)\left[1-f_1(E)\right]dE_\rho \tag{5.12}$$

where, E_ρ is longitudinal energy amount of total energy, E_{max} and E_{min} are the values of maximum and minimum energy respectively. In these expressions, the total energy, E_{Tot} has been split into a longitudinal energy part, E_ρ and a transversal energy part E_x. Therefore,

$$E_\rho = \frac{\hbar^2\left(k_y^2 + k_z^2\right)}{2m_{eff}} = \left(\frac{\hbar^2 k_\rho^2}{2m_{eff}}\right)$$
$$E_x = \left(\frac{\hbar^2 k_x^2}{2m_{eff}}\right) \tag{5.13}$$

Evaluating the difference of tunneling current density, $J_T = J_{T(1\to2)} - J_{T(2\to1)}$, the net current through the interface equals,

$$J_T = \left(\frac{4\pi m_{eff} q}{h^3}\right)\int_{E_{min}}^{E_{max}} T_C\left(E_x\right)d\left(E_x\right)\int_0^\infty \left(f_1\left(E\right) - f_2\left(E\right)\right)dE_\rho \tag{5.14}$$

This eq (5.14) is generally written by way of an integral above the product of two independent parts, which is only determined by the energy perpendicular to the interface: the transmission coefficient $T_C(E_x)$ and the supply function $N(E_x)$:

$$J_T = \left(\frac{4\pi m_{eff} q}{h^3}\right)\int_{E_{min}}^{E_{max}} T_C\left(E_x\right)N\left(E_x\right)dE_x \tag{5.15}$$

The values of $E_{\max}$ and $E_{\min}$ depend on the considered tunneling process: (i) ECB: $E_{\max}$ is the maximum conduction band edge of the metal and semiconductor, and $E_{\max}$ is the highest conduction band edge of the oxide layer, (ii) HVB: $E_{\max}$ is the absolute value of the lowermost valence band edge of the electrode and $E_{\max}$ is the absolute value of the lowest valence band edge of the oxide layer. The sign of the integration need to be altered, (iii) EVB: $E_{\max}$ is the lowermost conduction band edge of the metal and semiconductor; $E_{\max}$ is the uppermost valence band edge of the metal and semiconductor. It is necessary to verify if $E_{\min} < E_{\max}$.

5.3 SUPPLY FUNCTION MODELING

In equilibrium condition, the energy distribution function of electrons or holes is written by Fermi–Dirac statistics,

$$f(E)=\frac{1}{1+exp\left(\frac{E_{Tot}-E_F}{k_BT}\right)} \tag{5.16}$$

where, k_B is the Boltzmann constant, T is the temperature in Kelvin and E_F is the Fermi energy. The eq (5.16) can be derived by the statistical thermodynamics. Already we have $E_{Tot}=E_\rho+E_x$. Again we know the supply function describes the difference in supply of carriers at the interface of the oxide layer and it is given as,

$$N(E_x)=\int_0^\infty\left[f_1(E)-f_2(E)\right]dE_\rho \tag{5.17}$$

where $f_1(E)$ and $f_2(E)$ signify the energy distribution functions near the interfaces. For the reason that the particular nature of these distributions is usually not known, approximated natures are usually used. Moreover, it is assumed to be isotropic distributions. Now splitting the integral in eq (5.17), $N(E_x)=\lambda_1(E_x)-\lambda_2(E_x)$; therefore, the values of $\lambda_1(E_x)$ and $\lambda_2(E_x)$ become

$$\lambda_i=\int_0^\infty f_i(E)\,dE_\rho=\int_0^\infty\frac{1}{1+exp\left(\frac{E_x-E_\rho-E_{f,i}}{k_BT}\right)}dE_\rho \tag{5.18}$$

where, $i=1,2$ and $E_{f,i}$ is the Fermi energy of any i[th] material. This expression can be integrated analytically using $\int\left[1/\{1+\exp(x)\}\right]dx=ln\left[1/\{1+exp(-x)\}\right]+C$. So eq (5.18) can be evaluated as,

$$\lambda_i=k_BT\ln\left[1+exp\left(-\frac{E_x-E_{f,i}}{k_BT}\right)\right] \tag{5.19}$$

where, negative sign represents the n-type MOS structure. So the total supply function calculated from eq (5.17) becomes,

$$N(E_x) = (k_B T)\ln\left(\frac{1+exp\left(-\frac{E_x - E_{f,1}}{k_B T}\right)}{1+exp\left(-\frac{E_x - E_{f,2} - q\varphi_{ox}}{k_B T}\right)}\right) \tag{5.20}$$

where, φ_{ox} is the energy band offset value.

5.4 TRANSMISSION COEFFICIENT

The voltage through the MOS devices in terms of well-known voltages for the different regions has been calculated by potential energy profile in Figure 5.1 (Maity et al., 2017). Then applying time-dependent Schrödinger waves equation in every region as approached in Maity et al., (2017) we get

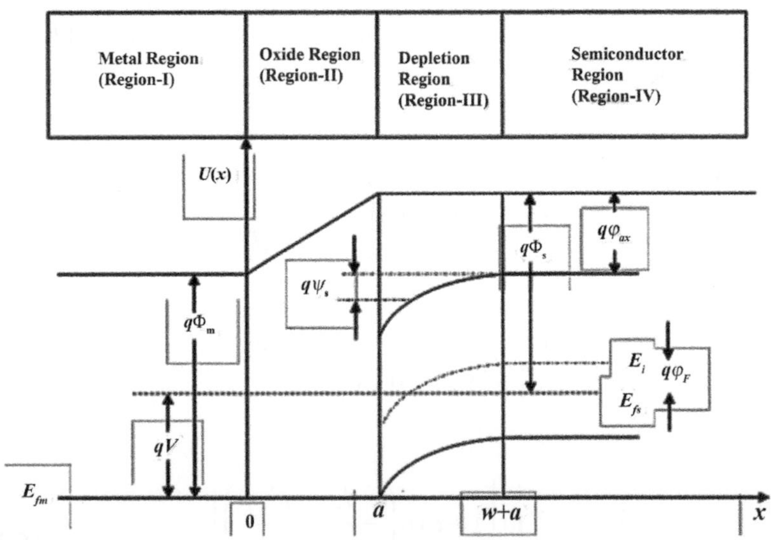

FIGURE 5.1 MOS structure with voltage profile.

Region I: Metal Region

$$\psi_m = Ae^{jk_1x} + Be^{-jk_1x} \tag{5.21}$$

Region II: Oxide Region

$$\psi_{ox}\left(U(x)\right) = CAi\left(U(x)\right) + DBi\left(U(x)\right) \tag{5.22}$$

Region III: Depletion Region

$$\psi_{sd} = \frac{Fe^{+\int\sqrt{(2m_{eff}/\hbar^2)(E-U(x))dx}} + Ge^{-\int\sqrt{(2m_{eff}/\hbar^2)(E-U(x))dx}}}{\left[(2m_{eff}/\hbar^2)(U(x)-E)\right]^{1/4}} \tag{5.23}$$

Region IV: Semiconductor Region

$$\psi_{sm} = He^{jk_4x} + Ie^{-jk_4x} \tag{5.24}$$

where A, B, C, D, F, G, H, and I are constants. $k_1 = \sqrt{(2m_{eff}/\hbar^2)(E_i - q\Phi_m)}$ and $k_4 = \left[\{2m_{eff}(E - q(\Phi_s + V - \varphi_{ox})\}/\hbar^2\right]^{1/2}$. Where, $U(x)$ is potential energy at distance x, Φ_s is work function of semiconductor, E is the total energy in metal, E_1 is electron energy, Φ_m work function of metal, ψ_m is potential wave function in metal region, ψ_{ox} is potential wave function in oxide region, ψ_{sd} is potential wave function in depletion region, ψ_{sm} is potential wave function in semiconductor region, *Bi* is Airy function of 1st kind and *Bi* is Airy function of 2nd kind. In this case we have considered time-independent Schrödinger wave equation; henceforth, the unknown coefficients of (5.21), (5.22), (5.23), and (5.24) are calculated using different boundary conditions for different regions of MOS devices (Maity et al., 2017). These are

$$\begin{aligned}
&\psi_m\big|_{x=0} = \psi_{ox}\big|_{x=0} \\
&\left(\frac{1}{m_{eff}}\right)\frac{d\psi_m}{dx}\bigg|_{x=0} = \left(\frac{1}{m_{ox}}\right)\frac{d\psi_{ox}}{dx}\bigg|_{x=0} \\
&\psi_{ox}\big|_{x=0} = \psi_{sd}\big|_{x=0} \\
&\left(\frac{1}{m_{ox}}\right)\frac{d\psi_{ox}}{dx}\bigg|_{x=a} = \left(\frac{1}{m_{eff}}\right)\frac{d\psi_{sd}}{dx}\bigg|_{x=a} \\
&\psi_{sd}\big|_{x=w+a} = \psi_{sm}\big|_{x=w+a} \\
&\left(\frac{1}{m_{eff}}\right)\frac{d\psi_{sd}}{dx}\bigg|_{x=w+a} = \left(\frac{1}{m_{eff}}\right)\frac{d\psi_{sm}}{dx}\bigg|_{x=w+a}
\end{aligned} \tag{5.25}$$

After solving the time-independent Schrödinger wave equation (Maity et al., 2017) and taking altogether the boundary conditions we can calculate the unidentified coefficients. To calculate them seven unknowns and simply six equations are available. Therefore, the finest that can be done is to find out the ratio of any two unidentified coefficients accurately. The transmittance coefficient can be defined by the ratio of transmission current with incidence current. The probability of density in region I and region IV can be derived as suggested by Maity et al. (2017) respectively,

$$J_1 = \frac{\hbar}{2jm_{eff}}\left[\psi_m^* \frac{d\psi_m}{dx} - \psi_m \frac{d\psi_m^*}{dx}\right] \tag{5.26}$$

$$J_4 = \frac{\hbar}{2jm_{eff}}\left[\psi_{sm}^* \frac{d\psi_{sm}}{dx} - \psi_{sm} \frac{d\psi_{sm}^*}{dx}\right] \tag{5.27}$$

where, $\psi_m = Ae^{ik_1x}$, $\psi_m^* = A^*e^{-ik_1x}$, $(d\psi_m/dx) = jk_1Ae^{ik_1x}$, $(d\psi_m^*/dx) = -jk_1A^*e^{-ik_1x}$, $\psi_{sm} = He^{ik_4x} + Ie^{-ik_4x}$, $\psi_{sm}^* = H^*e^{-ik_4x} + I^*e^{ik_4x}$, $(d\psi_s/dx) = jk_4He^{ik_4x} - jk_4Ie^{-ik_4x}$, and $(d\psi_s^*/dx) = -jk_4H^*e^{-ik_4x} + jk_4Ie^{ik_4x}$.

After solution as suggested by Maity et al., (2017) the transmission coefficient can be established as

$$T(E_x) = \frac{\left[\frac{\hbar k_1}{m_{eff}}\right]|A|^2}{\left[\frac{\hbar k_4}{m_{eff}}\right]|H|^2} = \left(\frac{k_1}{k_4}\right)\left|\frac{A}{H}\right|\left|\frac{A}{H}\right|^* \tag{5.28}$$

5.5 ANALYSIS AND APPLICATION TO HIGH-k MATERIAL

Figure 5.2 describes the tunneling current density values with the applied gate voltage. It illustrates the increase of total tunneling current density with increase of applied gate potential with the dissimilar oxide thicknesses, 1 nm and 1.2 nm. The current movement is enforced by the external voltage across the oxide layer of MOS devices (Maity et al., 2017).

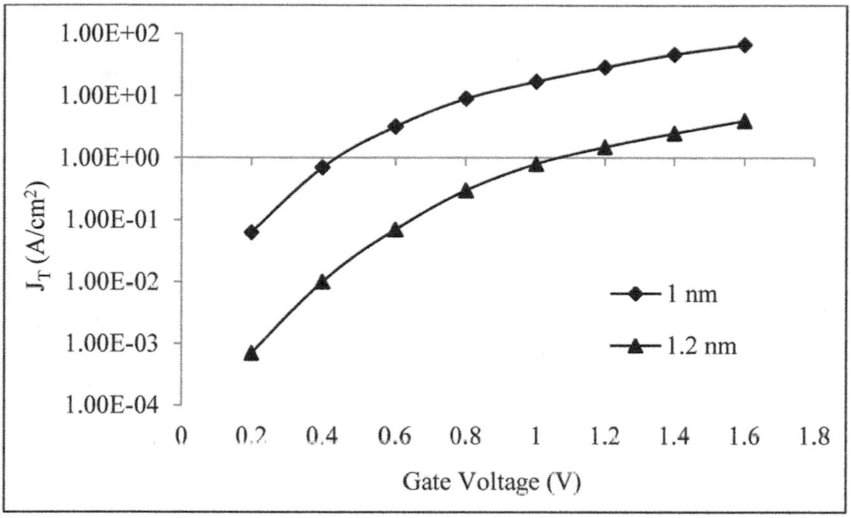

FIGURE 5.2 Tunneling current density with applied gate voltage for SiO_2.

The p-type semiconductor substrate-based MOS devices have been considered for the investigation comprising of SiO_2 as gate dielectric material. It is very clear to realize that the potential of gate terminal has a significant influence on the leakage current of MOS devices. While the gate oxide thickness is less than approximately 4 nm, the notable tunneling current occurrence is direct tunneling for a gate potential lesser than 2.5 V (Zhu and McNamara, 2015). The gate terminal current creates from the electrons tunneling from the semiconductor region to metal region for a positive gate potential supplied in the MOS devices. Electrons can tunnel from one or the other the conduction band that is recognized as ECB tunneling or the valence band is known as EVB tunneling (Zhu and McNamara, 2015). On the one hand it is recommended that at a lesser gate voltage values, the ECB tunneling current leads. On the other hand at the greater applied gate potential, the EVB tunneling current increases quickly. Then the total current density of a MOS structure is determined by both ECB tunneling current and EVB tunneling current (Zhu and McNamara, 2015). The tunneling current density with gate potential characteristic of high-k dielectric material HfO_2 is presented in Figure 5.3.

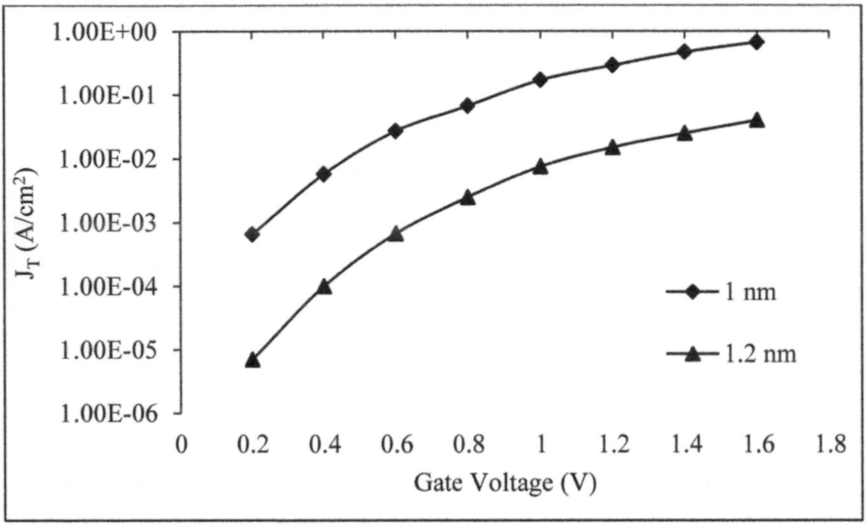

FIGURE 5.3 Tunneling current densities with applied gate voltages for HfO_2.

The physical characteristics of the diagrams are identical as that of SiO_2 material as a gate dielectric but due to the high-k dielectric material tunneling current is much lesser compared to SiO_2 material. In this case the investigation for the tunneling current density is constrained to the room temperatures, as a result any temperature effect involvement to the tunneling current association in the MOS devices can be ignored (Maity et al., 2017). As anticipated, gate tunneling current/leakage current increases near linearly with the decrease in gate oxide/insulator thickness as presented in Figure 5.4.

This provides poor performance of the transistor. At this time the applied gate potentials are 1 V and 0.8 V. For that reason the gate oxide thickness need to be cautiously selected to avoid the gate tunneling current. The variation of tunneling current density with the effective oxide thickness for outcomes of the model on behalf of high-k dielectric material HfO_2 is presented in Figure 5.5. The characteristics of the plots are identical as for SiO_2 and as expected when oxide thickness decreases the tunneling current increases and lesser tunneling current is observed for high-k material.

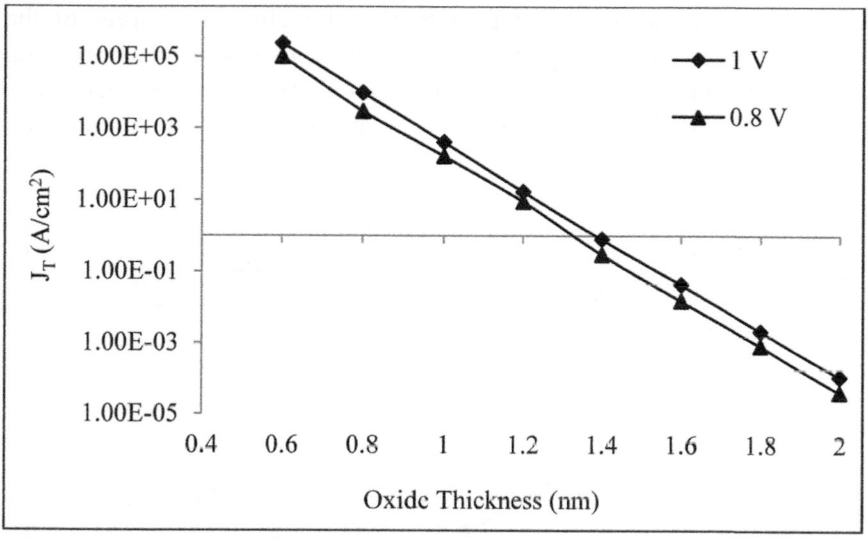

FIGURE 5.4 Tunneling current densities with oxide thickness for SiO_2.

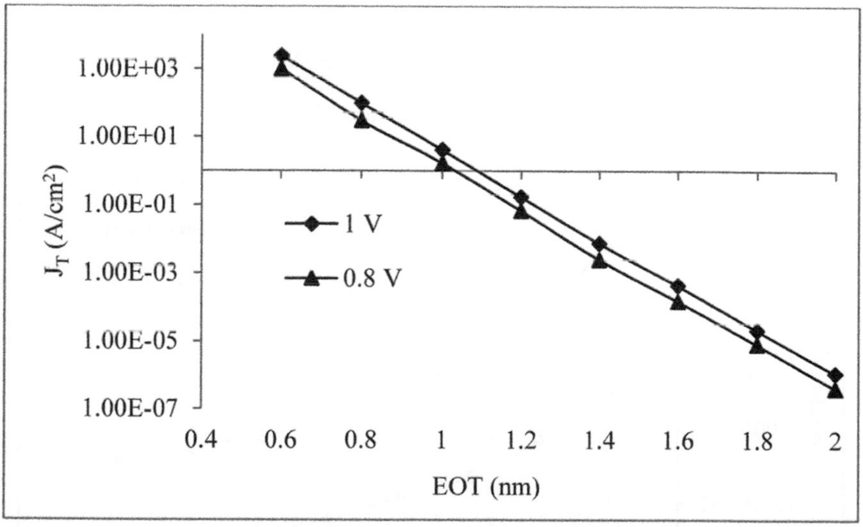

FIGURE 5.5 Tunneling current density with oxide thickness for HfO_2.

Figure 5.6 illustrates the tunnel resistivity variation as a function of applied gate potential for different oxide thicknesses, 1 nm and 1.2 nm as suggested by Maity et al., (2017; 2018; 2016). It checks that on behalf of lesser values of applied gate potential altogether the curves of the tunneling resistivity with applied gate voltage transfer toward a horizontal asymptote. For that reason the junctions exhibit ohmic characteristics (Maity et al., 2017). Tunnel resistivity is also evaluated for HfO_2 material, which is presented in Figure 5.7.

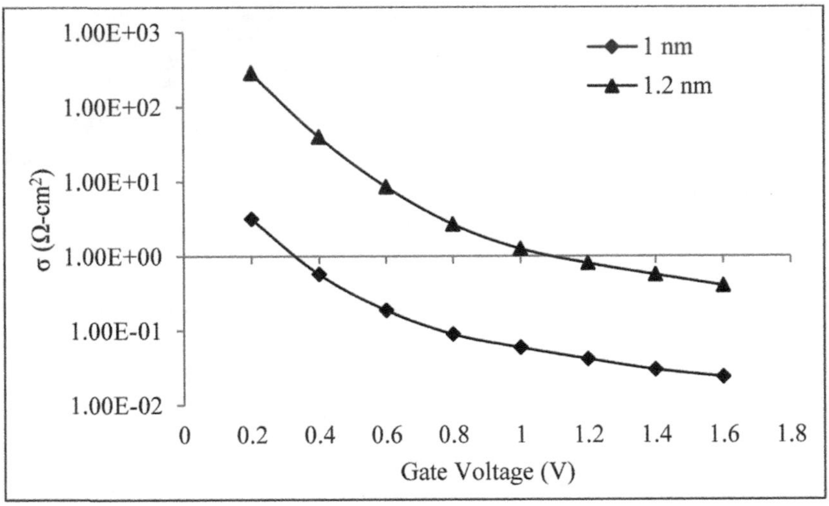

FIGURE 5.6 Tunneling resistivity with gate voltage of SiO_2.

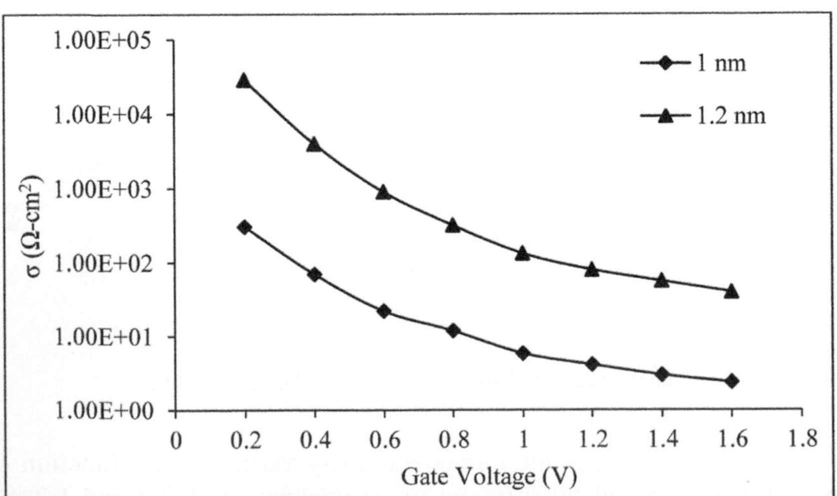

FIGURE 5.7 Tunneling resistivity with gate voltage of HfO_2.

5.6 SUMMARY

Based on the Tsu-Esaki model and considering potential energy profile for each region of the gate-dielectric-semiconductor structure, a tunneling current model has been developed to understand the behavior of the gate current as a function of applied potential, dielectric thickness, and temperature. A p-substrate silicon-based n-channel metal-oxide-semiconductor structure has been investigated comprising of SiO_2 material as the dielectric material along with promising high-k dielectric material HfO_2. Schrödinger's wave equation has been solved for the different potential energy regions for evaluation of transmission coefficients. The tunnel resistivity is also consistently evaluated using this tunneling model.

KEYWORDS

- **tunneling current**
- **tunneling resistivity**
- **transmission coefficient**
- **supply function model**
- **high-k materials**
- **hafnium oxide**

REFERENCES

Das, S. Transitioning from Microelectronics to Nanoelectronics. *IEEE Comp. Soc.* **2011,** *44*, 18–19.

Fu, C.; Liao, K.; Liu, L.; Li, C.; Chen, T.; Cheng, J.; Lu, C. An Ultralow EOT Ge MOS Device with Tetragonal HfO_2 and High Quality Hf_XGe_YO Interfacial Layer. *IEEE Trans. Electron Dev.* **2014,** *61*, 2662–2667.

Ganapathi, K. L.; Bhattacharjee, S.; Mohan, S.; Bhat, N. High Performance HfO_2 Back Gated Multilayer MoS_2 Transistor. *IEEE Electron Device Lett.* **2016,** *37*, 797–800.

Lin, Y.; Chian, C.; Linb, C.; Chang, C.; Lei, T. Novel Two-Bit HfO_2 Nanocrystal Nonvolatile Flash Memory. *IEEE Trans. Electron Dev.* **2006,** *53*, 782–789.

Litta, E.; Hellstrom, P.; Ostling, M. Threshold Voltage Control in TmSiO/HfO_2 High-k/ Metal Gate MOSFETs. *Solid-State Electron.* **2015,** *108*, 24–29.

Lo, S. H.; Buchanan, D. A; Taur, Y.; Wang, W. Quantum-Mechanical Modeling of Electron Tunneling Current from the Inversion Layer of Ultra-Thin-Oxide MOSFETs. *IEEE Electron Device Lett.* **1997,** *18*, 219–223.

Maity, N. P.; Maity, R.; Thapa, R. K.; Baishya, S. Study of Interface Charge Densities for ZrO_2 and HfO_2 Based Metal Oxide Semiconductor Devices. *Adv. Mater. Sci. Eng.* **2014,** *2014*, 1–6.

Maity, N. P.; Maity, R.; Thapa, R. K.; Baishya, S. Effect of Image Force on Tunneling Current for Ultra Thin Oxide Layer Based Metal Oxide Semiconductor Devices. *Nanosci. Nanotech. Lett.* **2015,** *7*, 331–333.

Maity, N. P.; Maity, R.; Thapa, R. K.; Baishya, S. A Tunneling Current Density Model for Ultra Thin HfO_2 High-k Dielectric Material Based MOS Devices. *Superlattices Microstruct.* **2016,** *95*, 24–32.

Maity, N. P.;Thakkur, R. R.; Maity, R.; Thapa, R. K.; Baishya, S. Analysis of Interface Charge Densities for High-k Dielectric Materials Based Metal-Oxide-Semiconductor Devices. *Int. J. Nanosci.* **2016,** *15*, 1660011 1–6.

Maity, N. P.; Maity, R.; Baishya, S. Voltage and Oxide Thickness Dependent Tunneling Current Density and Tunnel Resistivity Model: Application to High-k Material HfO_2 Based MOS Devices. *Superlattices Microstruct.* **2017,** *111*, 628–641.

Maity, N. P.; Maity, R.; Thapa, R. K.; Baishya, S. Influence of Image Force Effect on Tunneling Current Density for High-k Material ZrO_2ultra Thin Films Based MOS Devices. *J. Nanoelectron. Optoelectron.* **2017,** *12* (1), 67–71.

Maity, N. P.; Maity, R.; Baishya, S. A Tunneling Current Model with a Realistic Barrier for Ultra Thin High-k Dielectric ZrO_2 Material Based MOS Devices. *Silicon* **2018,** *10* (4) 1645–1652.

Maity, N. P.; Maity, R.; Baishya, S. An Analytical Model for the Surface Potential and Threshold Voltage of a Double-Gate Heterojunction Tunnel FinFET, *J. Comput. Electron.* **2018,** Online published (doi.org/10.1007/s10825-018-01279-5).

Maity, N. P.; Maity, R.; Maity, S.; Baishya, S. Comparative Analysis of the Quantum FinFET and Trigate FinFET Based on Modeling and Simulation, *J. Comput. Electron.* **2019,** Online published (doi.org/10.1007/s10825-018-01294-z).

Momose, H. S.; Ono, M.; Yoshitomi, T.; Ohguro, T.; Nakamura, S.; Saito, M.; Iwai, H. 1.5 nm Direct-Tunneling Gate Oxide Si MOSFET's. *IEEE Trans. Electron Dev.* **1996,** *3*, 1233–1242.

Sochacki, M.; Krol, K.; Waskiewicz, M.; Racka, K.; Szmidt, J. Interface Traps in Al/HfO_2/SiO_2/4H-SiC Metal-Insulator-Semiconductor (MIS) Structures Studied by the Thermally-Stimulated Current (TSC) Technique. *Microelectron. Eng.* **2016,** *157*, 46–51.

Taur, Y.; Buchanan, D. A.; Chen, W.; Frank, D. J.; Ismail K. E.; Lo, S. H.; Sai-Halasz, G. A.; Viswanathan, R. G.; Wann, H.; Wind, S.; Wong, H. S. CMOS Scaling into the Nanometer Regime. *In Proc. IEEE* **1997,** *85*, 486–504.

Yu, X.; Zhu, C.; Yu, M.; Kwong, D. Improvements on Surface Carrier Mobility and Electrical Stability of MOSFETs Using HfTaO Gate Dielectric. *IEEE Trans. Electron Dev.* **2004,** *51*, 2154–2160.

Zhu, L.; McNamara, S. Low Power Tunneling Current Strain Sensor Using MOS Capacitors. *IEEE J. Micromech. Sys.* **2015,** *24*, 755–762.

CHAPTER 6

Analysis of Interface Charge Density: Application to High-k Material Tantalum Pentoxide

N. P. Maity* and Reshmi Maity

Department of Electronics and Communication Engineering, Mizoram University (A Central University), Aizawl 796004, India

*Corresponding author. E-mail: maity_niladri@rediffmail.com

ABSTRACT

In this chapter a thickness-dependent interfacial distribution of oxide charges in MOS device structures has been analytically studied. The interface trap charges have been investigated and measured based on typical capacitance-voltage method (general method). The interface charge densities also have been investigated and measured based on conductance method. Linear increase in interface charge density has been observed as the dielectric constant of the oxide increases and comparable increase is also originated by decreasing the oxide thickness.

6.1 INTRODUCTION

In advanced semiconductor technology, ultimate oxide scaling down to equivalent oxide thickness (EOT) of approximately 0.5 nm is required, in order to achieve faster switching speed and higher device density (Houssa et al., 2006; Rathee et al., 2010). However, aggressive scaling of MOS transistors commanded to the replacement of conventional SiO_2 material and henceforth the integration of high-k dielectric materials into metal oxide semiconductor (MOS) devices. Nonetheless, replacing SiO_2 with high-k materials is a thoughtful challenge from fabrication point of view. This is because, although high-k materials compromise higher capacitance, they frequently suffer from poor electrical quality of dielectric-semiconductor interface and are repeatedly associated with lower dielectric breakdown voltages and decreased lifetimes (Maity et al., 2014; 2016; 2018; 2019; 2017).

In this chapter, the defects of semiconductor-oxide interface are broadly discussed for ultrathin MOS devices and the measurement techniques are also explained. Then interface trap charge has been evaluated by capacitance-voltage method and conductance method. The results have been validated with the device simulation tool. The influences of selected promising high-k dielectric material tantalum pentoxide (Ta_2O_5) on interface charge also have been carried out.

6.2 IDEAL MOS CAPACITOR

The capacitance of the MOS structure is the series combination of the oxide capacitance, C_{ox} and the semiconductor capacitance, C_s as shown in Figure 6.1.

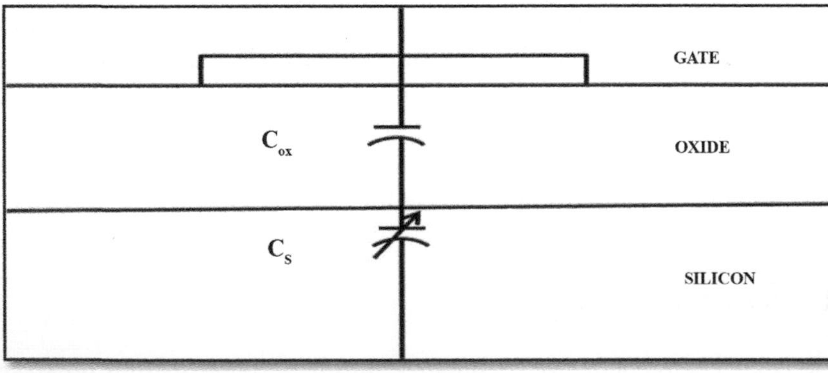

FIGURE 6.1 A simple circuit of MOS capacitor.

So, the semiconductor capacitance is $C_s = dQ_s/dV_s$. Where, V_s is the charge induced in the semiconductor under the oxide and V_s is the semiconductor potential. So, the total capacitance is given for a MOS device,

$$C_{mos} = \left[\frac{C_{ox}C_s}{C_{ox}+C_s}\right] \tag{6.1}$$

The three important regimes of accumulation, depletion, and inversion are reflected in capacitance-voltage characteristics. In the accumulation region (negative V_{Gs}), the holes accumulate at the surface and C_s is much larger than C_{ox}. This is because a small change in bias causes a large change in Q_s in the accumulation regime. The MOS capacitance is then $C_{mos} \cong C_{ox} = \varepsilon_{ox}/t_{ox}$. ε_{ox} is the permittivity of the oxide material. As the gate voltage V_G becomes positive and the channel is depleted of holes, the depletion capacitance becomes important. Under this condition the total capacitance is $C_{mos} = (C_{ox}C_s)/(C_{ox}+C_s)$. As the device gets increasingly depleted, the value of C_{mos} decreases. At the inversion condition, the depletion width reaches its maximum value (W_{max}). At this point there is essentially no free carrier density. If the bias is further increased, the free electron start to collect in the inversion regime and the depletion width remains unchanged with bias. The capacitance of the semiconductor again increases since a small change in V_s causes a large change in Q_s. The capacitance of the MOS device thus returns toward the value of C_{ox}. Therefore,

$$C_{mos(inversion)} \cong C_{ox} = \frac{\varepsilon_{ox}}{t_{ox}} \tag{6.2}$$

In the above-mentioned discussion we have assumed that there are no fixed charges in the oxide region. In reality, there are two types of charges that associate with oxide layer, namely, oxide charge and interface trap charge. The two charges have different effects on capacitance-voltage characteristics.

6.3 NONIDEAL MOS CAPACITOR

SiO_2 is often treated as an ideal insulator material, where there are no traps or states occur at the interface of Si-SiO_2. But in real devices, the Si-SiO_2 interface and bulk SiO_2 are distant from being electrically neutral. These may be instigated by positive or negative charges at the Si-SiO_2 interface or by mobile ionic charges and fixed charges trapped within the oxide region and itself, which are often generated during the fabrication process. The interface states (Kundu et al., 2011; Goetzberger and Sze, 1969; Deal,

1974; 1980; Schorder, 1990; Uren et al., 1992, Maity et al., 2014; 2016) are situated at or very neighboring to the semiconductor/oxide interface with energy distributed along the band gap of the semiconductor. Electrons or holes become trapped in these states and performance like charges at the interface (Nicollian and Brews, 2002). Hence, in reality, there are two types of charges that associate with oxide layer, namely: interface trap charge and oxide charge.

Figure 6.2 shows the interface trap charges and oxide charges behavior for a MOS device. The features that distinguish interface trap charge from oxide charge are that interface charge varies with gate bias whereas oxide charge is independent of gate bias. There are three types of oxide charges associated with the Si-SiO_2 system. They are fixed oxide charge, mobile oxide charge, and oxide trapped charge. All of these charges are greatly dependent on the semiconductor device fabrication process (Maity et al., 2014; 2016; Shanfield et al., 2002).

The easiest and most extensively used technique for determining oxide charge density is to infer this density from the voltage shift of a capacitance-voltage curve. A distinctive parameter of attention related to interface trap charge is the interface charge density, D_{it}, as a function of energy in the silicon band gap. Probable approaches for finding interface charge density and their relative merits are: capacitance-voltage method and conductance method. Capacitance-voltage method is the general method to calculate interface charge for a MOS structure. The conductance method is very popular and an accurate model for evaluation of interface charge density. It extracts interface trap level density, capture probability, and the time constant dispersion from the real component of the admittance. Recently, the conductance method has been explored and adapted for thoroughly proving the electrical passivation of novel semiconductor-dielectric interfaces (Lin et al., 2009; Martens et al., 2008; Cheng et al., 2011; Maity et al., 2014; 2016; Mamatrishat et al., 2012; Li et al., 2010; Srinivasan and Pandya, 2011).

6.4 DEFECTS AT Si-SiO_2 INTERFACE

6.4.1 *Interface Trapped Charges*

The interface trap can be positively or negatively charged trap at the oxide-semiconductor interface depending whether the trap is acceptor or donor category. An acceptor-type trap turns into negatively charged when it increases an additional electron and becomes neutral when it loses the additional electron. A donor-type trap turns into positively charged when it losses an electron and neutral when it recovers the lost electron. These kinds of charges have also been denoted to as surface states, fast states, or interface states. Under equilibrium circumstance, the tenancy of these interface traps is governed by the location of the Fermi level in the similar technique as for any additional electron energy level. Interface traps are normally initiated by trivalent silicon, which happens when silicon atoms bond to merely three oxygen atoms as a substitute of four. This type of defect is called amphoteric, which is owing to structural, oxidation-induced defects, metal impurities, incomplete bonds, and absorption of imported material at the silicon surface. The density of the interface trap charge can be condensed as soon as the surface is thermally oxidized. This characteristic of the Si-SiO_2 system is exceptional and is one of the most important causes for using thermally grown SiO_2 for gate oxide applications.

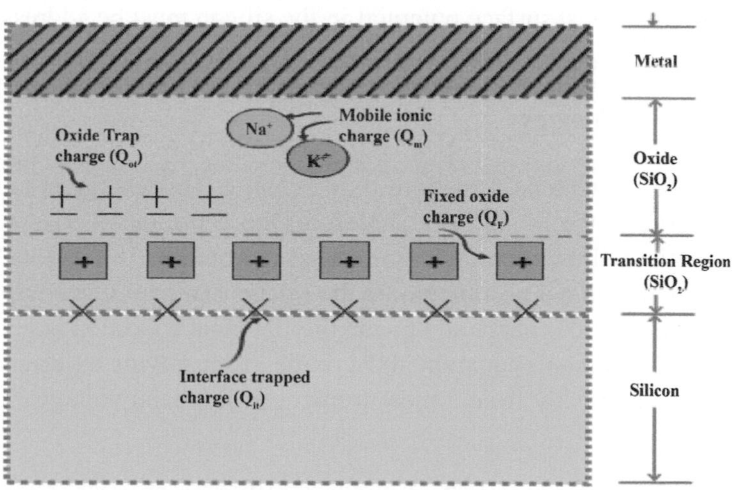

FIGURE 6.2 Interface charge and oxide charges.

Also interface trap is due to the Si-SiO_2 interface properties and depends on the chemical composition of this interface. The interface trap density is orientation dependent, for example, in <100> orientation the interface trap density is about an order of magnitude smaller than that in <111> orientation. A typical parameter of interest related to interface trap charge is the interface charge density, D_{it} as a function of energy in the silicon band gap. A low concentration or interface traps is indicative of a high quality interface between the oxide and semiconductor (Nicollian and Brews, 2002; Deal, 1980; Arora, 2007; Shanfield et al., 2002; Lu et al., 2011; Hurley et al., 2013; Schorder, 1990).

Possible methods for finding D_{it} and their relative merits are as follows:

1. Conductance-Voltage (G-V) Method: This is the most accurate technique developed by Nicollian and Brews (2002).
2. Comparison of an experimental quasi-static characteristic to an experimental 1 MHz characteristic. This produces data over a broader section of the band gap and eradicates the requisite for any theoretical comparison.
3. Comparison of an experimental quasi-static characteristic to the theoretical quasi-static characteristic. The consequences cover the broadest range in the band gap, nonetheless the concern of converting the plot of capacitance against gate voltage to capacitance against surface potential in the silicon must be addressed.

6.4.2 *Oxide Charges*

Let $n_0\,(x)$ be the volume density of oxide charge. This charge induces an image charge in the Si surface. Therefore, the capacitance at any given gate potential will be dissimilar from what it would be if $n_0\,(x) = 0$. The gate potential required to compensate the image charge produced in the Si surface by this oxide charge and create the flat band condition is given by integration of Poisson's equation in the oxide. After solving oxide charges, Q_0 is obtained directly from a measurement of flat band voltage V_{FB} and it gives

$$V_{FB} = \left[\Phi_{ms} - \left({Q_0}/{C_{ox}}\right)\right] \tag{6.3}$$

The equation shows that the entire capacitance-voltage curve is shifted along the voltage axis with respect to the ideal capacitance-voltage curve $\left[n_0\left(x\right)=0\right]$ by the amount $V_{FB} - \Phi_{ms}$ and the effects of Φ_{ms} and Q_o on Capacitance-Voltage characteristics of MOS capacitor is shown in Figure

6.3 (Nicollian and Brews, 2002; Deal, 1980; Arora, 2007; Shanfield et al., 2002; Lu et al., 2011; Maity et al., 2014; 2016; Hurley et al., 2013; Schorder, 1990).

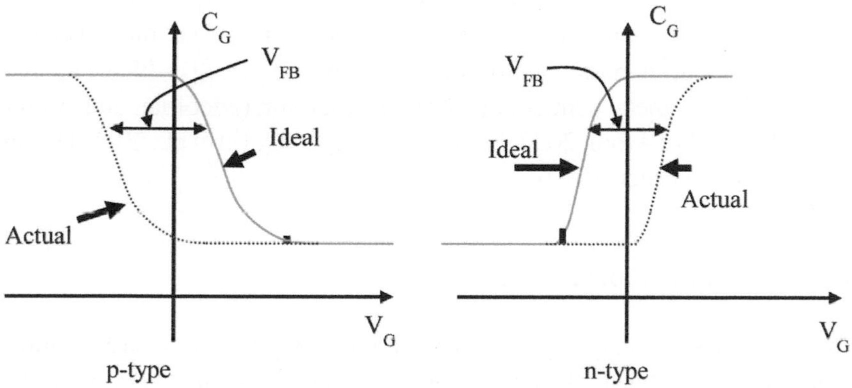

FIGURE 6.3 Effects of Φ_{ms} and Q_o on capacitance-voltage characteristics of MOS capacitor.

6.4.2.1 *Fixed Oxide Charges*

The fixed oxide charge, Q_f is a positive charge, primarily due to structural defects in the oxide layer < 25 Å from the Si-SiO_2 interface (located within approximately 3 nm of the Si-SiO_2 interface).The accurate character of the fixed oxide charges is not well-known. In general, these charges are immobile and cannot be charged or discharged over a broad variation of surface potential and are independent of the gate potential. It is affected by temperatures above 500°C and by the ambient atmosphere but is not affected by the oxide thickness. Conversely, it increases when the structure is exposed to high energy radiation. Moreover, Q_f depends on oxidation and annealing conditions and on Si orientation. The density of Q_f is highly dependent on the process used to generate the oxide layer and on the orientation of the Si. Usually it has been observed that when the oxidation is stopped, some ionic silicon is left near the interface and it may result in the positive interface charge Q_f. Typical values for Q_f are in the order of 10^{10}–10^{11}cm^{-2}, depending on the process environment and typical fixed-oxide charge densities for a cautiously treated Si-SiO_2 interface system are about 10^{10} cm^{-2} for <100>surface and about 5×10^{10} cm^{-2} for a <111>surface.

Q_f is determined by comparing the flat band voltage shift of an experimental capacitance-voltage curve with a theoretical curve and measureing the voltage shift. To determine Q_f, one should eradicate or at least reduce the effects of all other oxide charges Q_0 and reduce the interface trapped charge to as low a value as possible. So, $Q_0 = Q_f$. Now we get, $Q_f = (\Phi_{ms} - V_{FB})C_{ox}$, where, Φ_{ms} is metal semiconductor work function (Nicollian and Brews, 2002; Deal, 1980; Arora, 2007; Shanfield et al., 2002; Lu et al., 2011; Hurley et al., 2013; Schorder, 1990).

6.4.2.2 Mobile Ionic Charges

The mobile oxide charge, Q_m is one of the most significant charge component in the insulator. It is due to ionized impurities such as sodium (Na^+) and, to a less amount, potassium (K^+) and lithium (Li^+). The alkali ions are difficult to control because they become mobile at high temperatures under an applied electric field. Seeing as, the mobile ionic charges are movable within the oxide layer under lift up temperature (e.g., >100° C) and high electric field operation, it causes stability difficulty in device. These positively charged ions could travel from the bulk of the insulator layer to the Si–SiO_2 interface over a period of time, gradually increasing the oxide charge there.

Mobile oxide charge is supposed to be typically sodium ions. Mobile ions can be detected by observing V_{FB} and threshold voltage shift under a gate potential at an eminent temperature (e.g., at 200° C) due to the movement of the ions in the insulator. Sodium contamination must be eliminated from the water, chemicals, and containers used in an MOS fabrication line in order to prevent instabilities in V_{FB} and V_T. The bias temperature stress (BTS) technique is one of the two techniques to control the mobile charge. Normally, MOS device is heated to 150°C to 250°C, and a gate potential to create an oxide electric field of around 106 V/cm is applied for 5–10 min for the charge to drift to one oxide interface. The transistor is then cooled to room temperature under bias and a capacitance-voltage curve is measured. The process is then repeated with the opposite bias polarity. Then the mobile charge can be measured from the flat band voltage shift, according to the following equation,

$$Q_m = \left[\left(C_{ox}/A\right)\Delta V_{FB}\right] \tag{6.4}$$

Where, $\Delta V_{FB} = V_{FB}^{-} - V_{FB}^{+}$, flat band voltage shift, V_{FB}^{+} represents flat band voltage shift correspond from positive to negative polarity bias, V_{FB}^{-} represents flat band voltage shift correspond from negative to positive polarity bias, and A corresponds to the area of test structure (Nicollian and Brews, 2002; Deal, 1980; Arora, 2007; Shanfield et al., 2002; Lu et al., 2011; Hurley et al., 2013; Schorder, 1990).

6.4.2.3 Oxide Trapped Charge

The oxide trap charge, Q_{ot} is owing to deficiencies throughout the bulk region of the insulator layer. This charge may be positive or negative depending whether holes or electrons are trapped in the bulk of the oxide. These charges can be generated, for example, by X-ray radiation or high-energy electron bombardment. The majority of process-related oxide trapped charge can be removed by low-temperature annealing method. Trapping might result from ionizing radiation, or supplementary comparable procedures and the V_T can shift in either direction. The oxide trapped charges get up from defects in the oxide region. These imperfections may be structural, chemical, or impurity associated. The defects, originally neutral, apprehension electrons or holes and turn into negatively or positively charged. In the meantime, very slight current flows through the oxide layer throughout typical device operation, the traps generally remain neutral. Conversely, if carriers are injected into the oxide layer, or ionizing radiation movements through the oxide, these traps can become charged. Both Q_m and Q_{ot} are distributed randomly throughout the oxide layer (Nicollian and Brews, 2002; Deal, 1980; Arora, 2007; Shanfield et al., 2002; Lu et al., 2011; Hurley et al., 2013; Schorder, 1990).

6.5 EVALUATION OF INTERFACE CHARGE USING C-V METHOD

6.5.1 Theoretical Analysis

The capacitance-voltage technique depends on the circumstance that the thickness of a reverse biased condition space charge area of a semiconductor junction device is determined by the applied voltage at gate terminal. This space charge area thickness requirement on voltage lies at the heart of the capacitance-voltage technique. The occurrence of the interface at the silicon surface acquaint with a clear perturbation to the periodic crystal structure of the semiconductor material and reasons some silicon-silicon bonds to be unfulfilled or "dangling". By way of a consequence there are energy states in the band gap at the silicon surface. These states are called "interface states" or "interface traps" (Wilk et al., 2001; Tapajna et al., 2006). Three insulator-associated parameters characteristically determined with capacitance-voltage measurement are: Oxide charge density, ρ_{ox}, Interface trap density, Φ_{ms}, and Gate semiconductor work function difference, Φ_{ms}. They are determined from the flat band voltage,

$$V_{FB} = \left[\Phi_{ms} - \left(1/(\varepsilon_{ox}\varepsilon_o)\right)\int_0^{t_{ox}} x\rho(x)dx\right] \tag{6.5}$$

Q_f and Q_{it} are assumed to be located at the Si-SiO_2 interface and the remaining mobile ionic and oxide trap charges in the SiO_2, leading to the V_{FB} expression,

$$V_{FB} = \Phi_{ms} - \left[\left\{\frac{1}{\varepsilon_{ox}\varepsilon_o}\right\}\int_0^{t_{ox}} xQ_f\delta(t_{ox})dx\right] - \left[\left\{\frac{1}{\varepsilon_{ox}\varepsilon_o}\right\}\int_0^{t_{ox}} x\rho_{ox}(x)dx\right] - \left[\left\{\frac{1}{\varepsilon_{ox}\varepsilon_o}\right\}\int_0^{t_{ox}} xQ_{it}(\phi_s)\delta(t_{ox})dx\right] \tag{6.6}$$

Where, δ is delta function and Q_{it} is a function of surface potential, ϕ_s. So, for uniform oxide charge density as specified by Nicollian and Brews (2002),

$$V_{FB} = \left[\Phi_{ms} - \left\{\frac{Q_f + Q_{it}(\phi_s)}{\varepsilon_{ox}\varepsilon_o}\right\}t_{ox} - \left\{\frac{\rho_{ox}}{2\varepsilon_{ox}\varepsilon_o}\right\}t_{ox}^2\right] \tag{6.7}$$

Soon supposing that only Q_{it} is existent in oxide layer,

$$V_{FB} = \Phi_{ms} - \left[\left\{\frac{Q_{it}(\phi_s)}{\varepsilon_{ox}\varepsilon_o}\right\}t_{ox}\right] \tag{6.8}$$

In the high frequency capacitance technique, capacitance value is measured as a function of gate potential with frequency fixed at higher sufficient value so that interface traps do not respond. At high frequency the total capacitance is given with $C_{it}(\omega)=0$ (because ω is excessively large value for any AC response of interface traps). That is, capacitance at high frequencies C_{HF} is specified by Nicollian and Brews (2002),

$$C_{HF} = \left[\frac{(C_s C_{ox})}{(C_s + C_{ox})}\right] \tag{6.9}$$

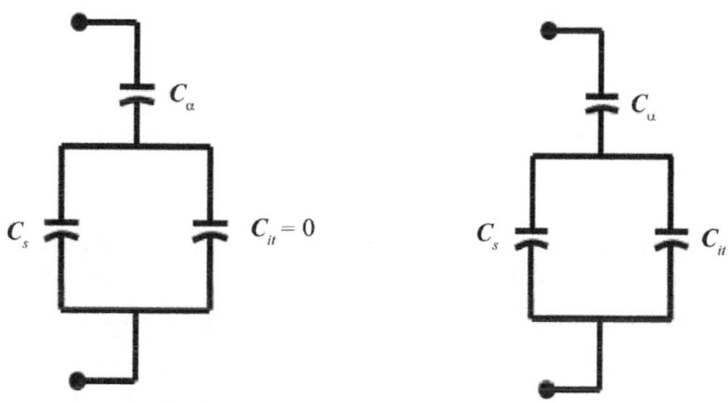

FIGURE 6.4 Equivalent MOS capacitor at (a) High frequency (b) Low frequency (Nicollian and Brews, 2002).

The high frequency capacitance of the MOS capacitor will be similar as that of an ultimate one without interface traps, on condition that C_s is identical. But now a frequency so low that interface trap response is instantaneous, that is, capacitance at low frequencies C_{LF} is given by Nicollian and Brews (2002),

$$\left[\frac{1}{C_{LF}}\right] = \left[\left\{\frac{1}{C_{ox}}\right\} + \left\{\frac{1}{C_s + C_{it}}\right\}\right] \tag{6.10}$$

Where, C_{LF} is the low frequency capacitance measured at gate bias V_G. Equivalent circuit for a MOS capacitor is shown in Figure 6.4. Now solving (6.10) for C_{it} yields, $C_{it} = \left[(1/C_{LF}) - (1/C_{ox})\right]^{-1} - C_S$, where C_s can be measured in strong accumulation condition. Using (6.8) C_s can be written as, $C_S = \left[(1/C_{HF}) - (1/C_{ox})\right]^{-1}$. Therefore, C_{it} is given by

$$C_{it} = \left(\frac{1}{C_{LF}} - \frac{1}{C_{ox}}\right)^{-1} - \left(\frac{1}{C_{HF}} - \frac{1}{C_{ox}}\right)^{-1} \tag{6.11}$$

Interface charge densities, D_{it} can be measured by $D_{it} = \left[C_{it} / q^2 \right]$. Where, q is the electronic charge. So, D_{it} can be expressed from the (6.11) as

$$D_{it} = \left(\frac{1}{q^2}\right)\left[\left\{\left(\frac{1}{C_{LF}} - \frac{1}{C_{ox}}\right)^{-1} - \left(\frac{1}{C_{HF}} - \frac{1}{C_{ox}}\right)^{-1}\right\}\right] \tag{6.12}$$

6.5.2 *Results*

The p-type substrate-based MOS devices have been investigated involving of SiO_2 along with high-k dielectric material, Ta_2O_5. Out of the several high-k dielectric materials, Ta_2O_5 is one of the encouraging materials for these circumstances. Ta_2O_5 has been previously considered for gigabit dynamic random access memory (DRAM) application. Ta_2O_5 has also been recommended for very high speed electronic devices. Pure Ta_2O_5 has very high storage capability (Georgievska et al., 2012). These materials are considered for different insulator thickness at p-type doping level of 1×10^{17} cm^{-3}. The variations of magnitude of the flat band voltage as a function of insulator thickness for SiO_2 and high-k material are shown in Figures 6.5 and 6.6 respectively. These graphs clearly demonstrated the validity of our assumption in eq (6.8).

Q_{it} can be extracted from the slope of plot between flat band voltages and oxide thickness. Subsequently, occurrence of lone interface charge at the oxide-semiconductor interface the value of Q_{it} has been calculated theoretically and through technology computer-aided design (TCAD) simulation. The theoretical value of the interface charge, Q_{it} present in Table 6.1 and the device simulation result obtained is presented in Table 6.2, which is increasing by dielectric constant value of materials.

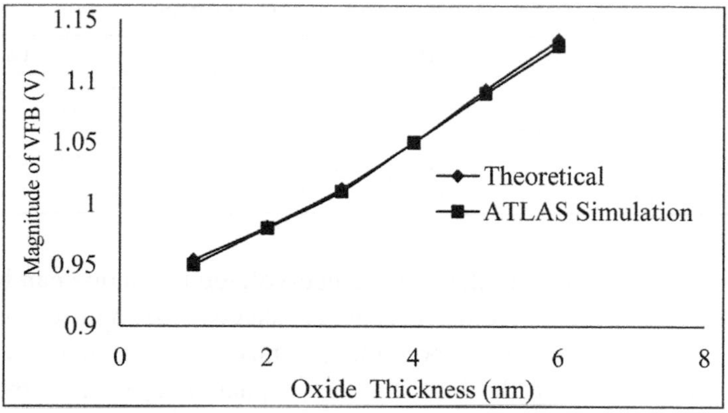

FIGURE 6.5 Variation of V_{FB} with oxide thickness for SiO_2.

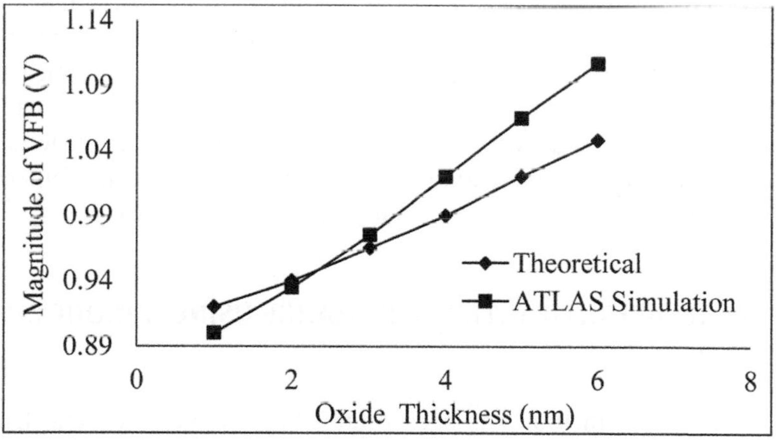

FIGURE 6.6 Variation of V_{FB} with oxide thickness for Ta_2O_5.

TABLE 6.1 Interface Charge Q_{it} in SiO_2, and Ta_2O_5 (Theoretical).

Material	Q_{it1}	Q_{it2}	Q_{it3}	Q_{it4}	Q_{it5}	Q_{it6}	Q_{it} (Average)
SiO_2	5.035×10^{10}	4.963×10^{10}	5.395×10^{10}	4.855×10^{10}	5.179×10^{10}	4.532×10^{10}	4.993×10^{10}
Ta_2O_5	13.79×10^{10}	14.00×10^{10}	15.82×10^{10}	12.78×10^{10}	13.39×10^{10}	12.17×10^{10}	13.66×10^{10}

TABLE 6.2 Interface Charge Q_{it} in SiO_2, and Ta_2O_5 (Simulation).

Material	Q_{it1}	Q_{it2}	Q_{it3}	Q_{it4}	Q_{it5}	Q_{it6}	$Q_{it\ (Average)}$
SiO_2	5.039×10^{10}	4.973×10^{10}	5.295×10^{10}	4.955×10^{10}	5.279×10^{10}	4.232×10^{10}	4.962×10^{10}
Ta_2O_5	13.797×10^{10}	14.002×10^{10}	14.821×10^{10}	12.787×10^{10}	13.390×10^{10}	12.172×10^{10}	13.594×10^{10}

As we have presented, the capacitance-voltage technique can be used to evaluate the interface-trapped charges because the input capacitance contains similar information about the interface traps. Difficulty arises in the technique, because the interface-trap capacitance must be extracted from the measured capacitance, which comprises of oxide capacitance, depletion layer capacitance, and interface-trap capacitance. As earlier pointed out, the capacitance value as function of applied voltage and frequency comprises of indistinguishable information about interface traps. Problems arise in taking out this information from the measured capacitance value. For this reason, the dissimilarity between two capacitance values is needed to be calculated. Accordingly, the capacitance-voltage method can be avoided on behalf of the measurement of interface charge densities of MOS devices (Maity et al., 2014; 2016).

6.6 EVALUATION OF INTERFACE CHARGE USING CONDUCTANCE METHOD

The conductance method has been established by Nicollian and Brews (2002) for calculating the oxide-semiconductor interface charge density, D_{it}. This technique is simplest in depletion because minority carrier effects are not important in this case. That is, in depletion interface trap occupancy alterations by capture and emission of majority carriers (holes in p-type and electrons in n-type). Minority carrier capture is unimportant because the minority carrier density is very low in depletion for biases not too close to mid gap. Similarly, minority carrier emission is negligible because the small-signal measurement disturbs the equilibrium only slightly and the emission rate is comparable to negligible capture rate. The conductance method extracts: interface trap level density, capture probability, and the time constant dispersion from the real component of the admittance.

Generally, in the conductance technique, interface trap levels are recognized over and done with the loss resulting from changes in their occupancy formed by minor variation of gate voltage. There will be an energy loss on together half of the ac cycle that must be provided by the signal source. This energy loss is measured as an equivalent parallel conductance G_p. In addition to an energy loss associated with capture and emission, interface trap also can hold an electron for some time after capture, that is, interface trap store charge. Therefore, there will be a capacitance C_{it} proportional to interface trap level density (Nicollian and Brews, 2002). Recently, the conductance method has been investigated and adapted for rigorously proving the electrical passivation of novel semiconductor-dielectric interfaces (Lin et al., 2009; Martens et al., 2008; Cheng et al., 2011; Mamatrishat et al, 2012; Maity et al., 2014; 2016).

6.6.1 *Theoretical Analysis*

The modeling of interface charge densities measurement has been done based on conductance method. Figure 6.7 shows an equivalent circuit model of metal oxide semiconductor capacitor. Figure 6.7(a) consists of C_{ox}, semiconductor capacitance per unit area, C_s, interface trap capacitance per unit area, C_{it} and interface trap resistance per unit area, R_{it}. For the MOS interface charge analysis it is appropriate to substitute the circuit of Figure 6.5(a) with Figure 6.7(b): Here C_P is the equivalent parallel capacitance and the G_P is the equivalent parallel conductance. So the admittance of Figure 6.7(a) and Figure 6.7(b) is given by (Maity et al., 2014),

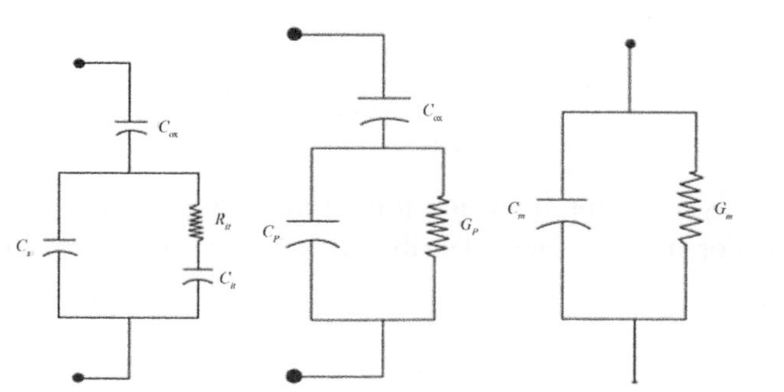

FIGURE 6.7 Equivalent circuits for conductance measurements: (a) MOS-*C* with interface trap time constant $\tau_{it} = C_{it}R_{it}$, (b) simplified circuit of (a), and (c) measured circuit (Reprinted from Maity et al., 2014).

$$Y_a = j\omega C_s + \left[\frac{1}{R_{it} + \frac{1}{j\omega C_{it}}} \right] = j\omega C_s + \left[\frac{j\omega C_{it}}{1 + j\omega\tau_{it}} \right] \tag{6.13}$$

Where, $\tau_{it} = C_{it}R_{it}$, then Multiplying both sides by $1 - j\omega\tau_{it}$ we get,

$$Y_a = j\omega C_s + \left[\left\{ \frac{j\omega q D_{it}}{1 + j\omega\tau_{it}} \right\} \left\{ \frac{1 - j\omega\tau_{it}}{1 - j\omega\tau_{it}} \right\} \right] = j\omega C_s + \left[\frac{\{ j\omega q D_{it} (1 - j\omega\tau_{it}) \}}{\{ 1 + (\omega\tau_{it})^2 \}} \right] \tag{6.14}$$

$$Y_a = j\omega C_s + \left[\frac{\{ j\omega q D_{it} (1 - j\omega\tau_{it}) \}}{\{ 1 + (\omega\tau_{it})^2 \}} \right] \tag{6.15}$$

$$Y_b = [j\omega C_P + G_P] \tag{6.16}$$

where, $C_{it} = qD_{it}$, $\omega = 2\pi f$, f is the measured frequency, q is the magnitude of electronic charge. The real part and imaginary part are compared, respectively, then we obtain,

$$C_P = C_S + \left[\frac{qD_{it}}{\{ 1 + (\omega\tau_{it})^2 \}} \right] \tag{6.17}$$

$$\left[\frac{G_P}{\omega} \right] = \left[\frac{q\omega\tau_{it}D_{it}}{\{ 1 + (\omega\tau_{it})^2 \}} \right] \tag{6.18}$$

This analysis is for interface trap with solitary energy level in the band gap. For uninterrupted distributed interface trap refer Shanfield et al. (2002),

$$C_P = C_S + \left[\left\{ \frac{qD_{it}}{\omega\tau_{it}} \right\} \tan^{-1}(\omega\tau_{it}) \right] \tag{6.19}$$

$$\left[\frac{G_P}{\omega}\right]=\left[\left\{\frac{qD_{it}}{2\omega\tau_{it}}\right\}\ln\left\{1+\left(\omega\tau_{it}\right)^2\right\}\right] \tag{6.20}$$

Dividing G_P by ω makes it symmetrical in $\omega\tau_{it}$ and G_P/ω is directly related to D_{it}. It is more operational to calculate D_{it}, because it has only a parameter related to interface trap without including C_s. For continuously distributed interface charge, we must consider Figure 6.7 (c), comprising of the measured equivalent parallel conductance G_m and measured capacitance C_m. Assuming negligible series resistance G_P/ω is given (Nicollian and Brews, 2002; Deal, 1980; Arora, 2007; Maity et al., 2014; 2016; Shanfield et al., 2002; Lu et al., 2011),

$$\left[\frac{G_P}{\omega}\right]=\left[\frac{\omega G_m C_{ox}^2}{\left\{G_m^2+\omega^2\left(C_{ox}-C_m\right)^2\right\}}\right] \tag{6.21}$$

D_{it} can be calculated from the obtained G_P/ω vs. f graph from eq (6.8), that is,

$$D_{it}=\left[\left\{\frac{1+\left(\omega\tau_{it}\right)^2}{q\omega\tau_{it}}\right\}\left\{\frac{G_P}{\omega}\right\}\right] \tag{6.22}$$

At maximum G_P/ω, the ω is the inverse of τ_{it}. In this case, the estimated expression of D_{it} can therefore be assumed in terms of the measured maximum conductance as suggested by Maity et al. (2014),

$$D_{it}=\left[\left\{\left(2/q\right)\left(G_P/\omega\right)\right\}_{\max}\right] \tag{6.23}$$

6.6.2 Results

The value of D_{it} has been calculated using conductance method consisting of SiO_2 material along with the high-k dielectric material, Ta_2O_5. It is a technique that substitutes MOS capacitor with corresponding circuit model and determines G_P/ω. The conductance of MOS has been measured from the conductance-voltage curve. Using the simulated capacitance-voltage and conductance-voltage characteristics curve by 2D device simulator, ATLAS of Silvaco TCAD for a wide variation of oxide thickness and biasing condition having doping concentration of $1 \times 10^{17}\ cm^{-3}$, the values of the C_m and G_m have been measured. The values of the C_m have been measured for different oxide thicknesses (1 nm to 3 nm) and the variation with frequency for SiO_2 material and the selected high-k dielectric material, Ta_2O_5 have been presented in Figures 6.8 and 6.9 respectively.

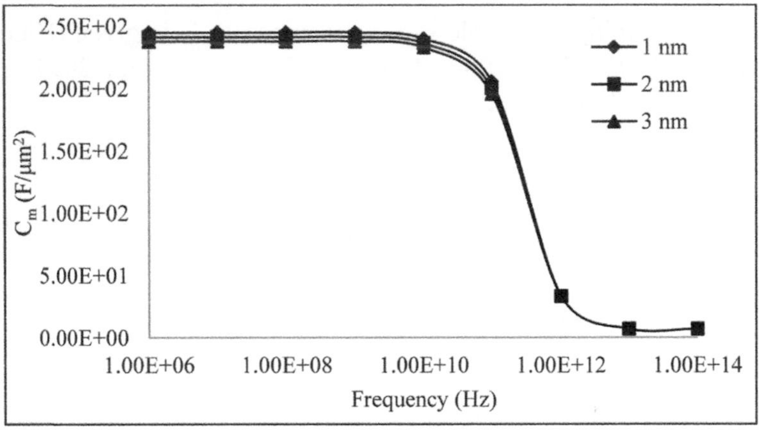

FIGURE 6.8 Variation of C_m with frequency for SiO_2.

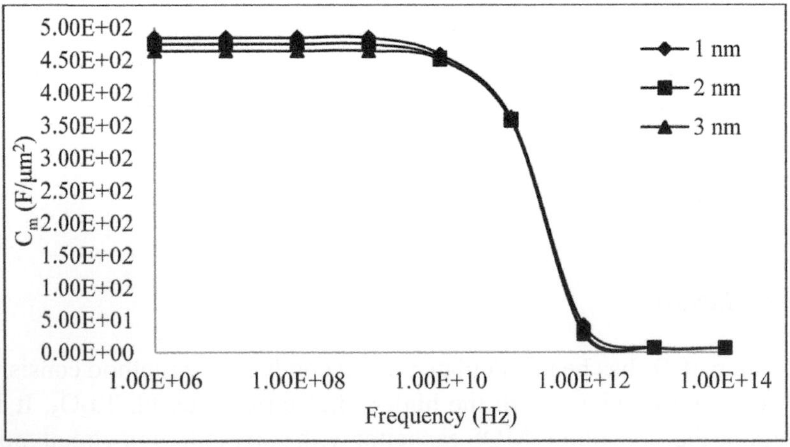

FIGURE 6.9 Variation of C_m with frequency for Ta_2O_5.

Similarly, the values of the G_m have been also measured for different oxide thicknesses (1 nm to 3 nm) and the variation with frequency for SiO_2 material and the selected high-k dielectric material have been shown in Figures 6.10 and F 6.11 respectively.

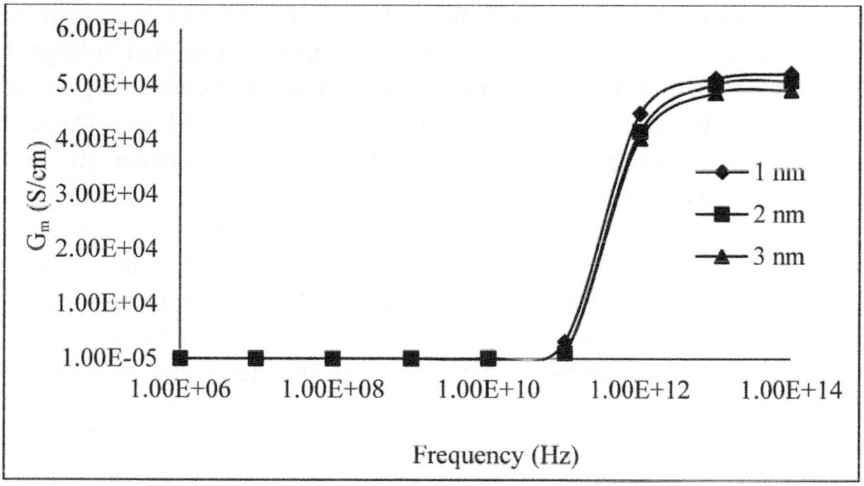

FIGURE 6.10 Variation of G_m with frequency for SiO_2.

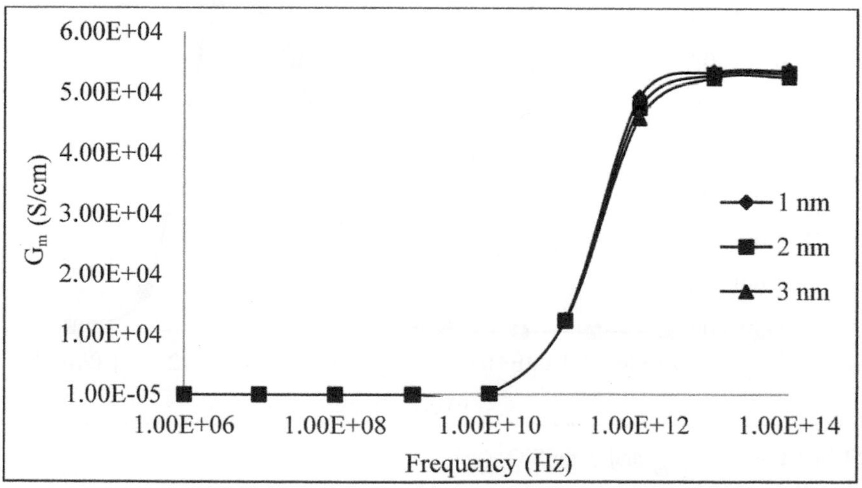

FIGURE 6.11 Variation of G_m with frequency for Ta_2O_5.

Figures 6.12 and 6.13 show the relationship of G_P/ω vs. f of SiO_2 and Ta_2O_5. These figures illustrate that the local maximum of every single curve specifies the magnitude of D_{it}. The peak points of the characteristics compute the interface state energy levels. Together G_P/ω and G_P will peak as a function of applied gate voltage but then again only G_P/ω will peak as a function of frequency. G_P/ω will highest at a lower frequency as soon as admittance is measured as a function of frequency with applied voltage at the gate terminal as parameter or at the gate potential neighboring to flat bands then when admittance is measured as a function of bias voltage at the gate with frequency as parameter. G_P/ω is maximum between 10^{11} and 10^{12} Hz for SiO_2 and 10^{10} for Ta_2O_5.

Figures 6.14 and 6.15 illustrate the relationship between D_{it} and f for SiO_2 and the high-k dielectric material Ta_2O_5 respectively. The value of D_{it} has been calculated at different frequencies based on most accurate technique conductance method. At smaller frequencies D_{it} has a constant and high value. As soon as the frequency touches the inverse of interface trap time constant, D_{it} starts reducing and at the very high frequencies the value goes to a saturation region of the MOS devices.

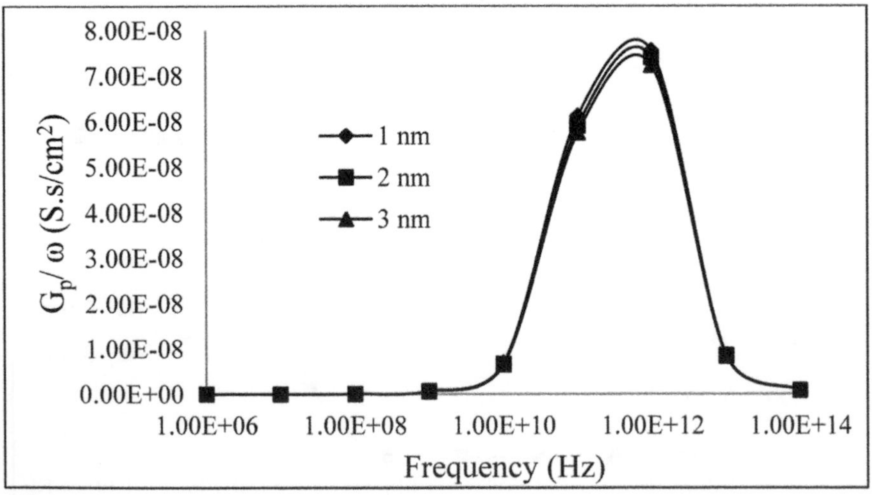

FIGURE 6.12 G_P/ω and f for SiO_2.

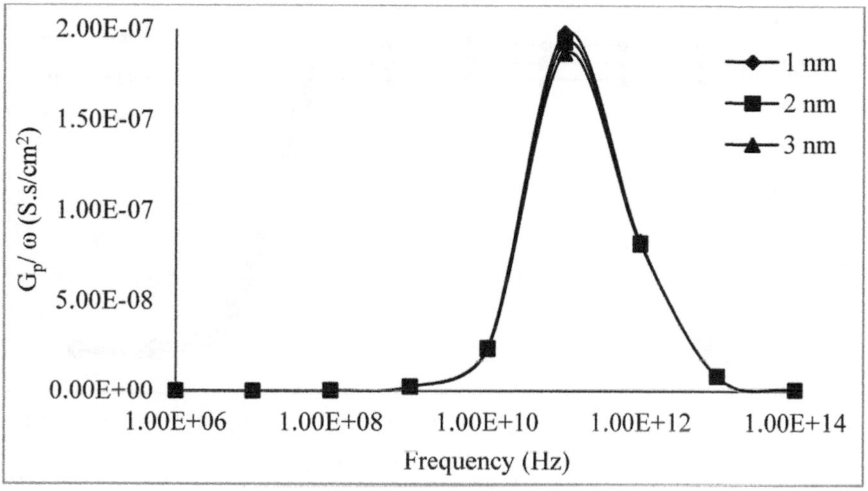

FIGURE 6.13 G_P/ω and f for Ta_2O_5.

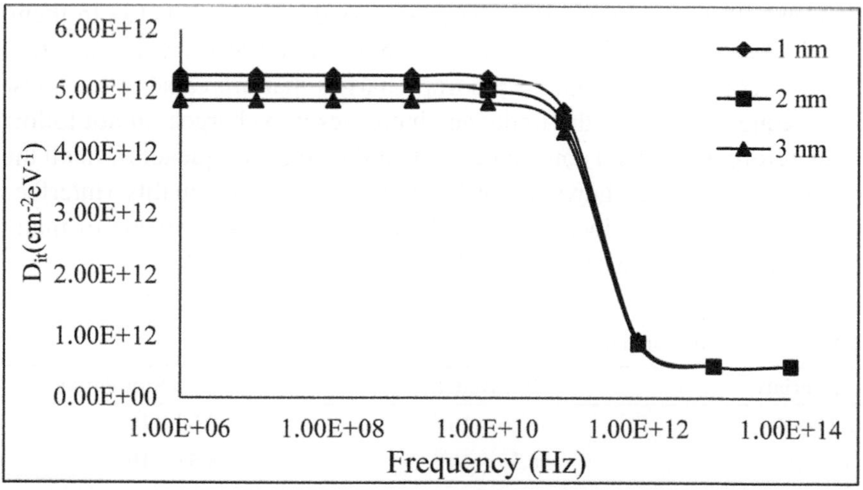

FIGURE 6.14 D_{it} and f for SiO_2 material at different oxide thicknesses.

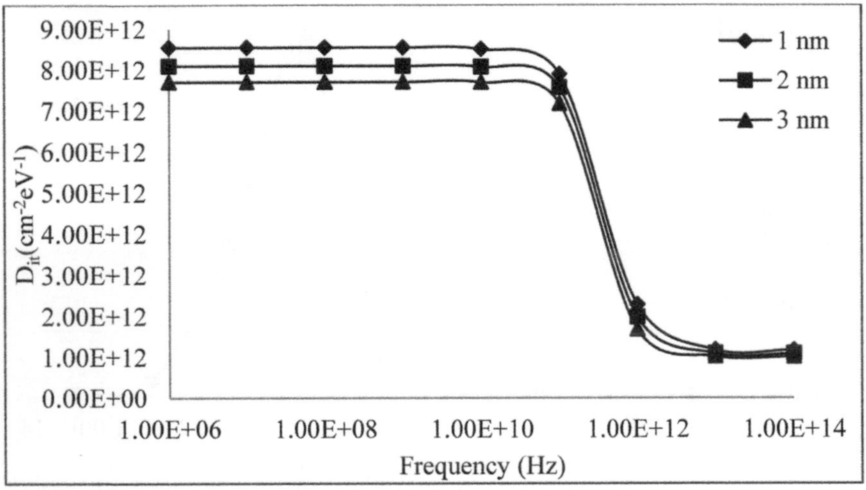

FIGURE 6.15 D_{it} and f for Ta_2O_5 material at different oxide thicknesses.

Table 6.3 indicates the interface charge densities value for analytical model and TCAD ATLAS simulation results at 1 nm of oxide thickness for SiO_2 and Ta_2O_5 materials. It illustrates a good agreement between them. The higher value of D_{it} at lower frequencies and a lower value of D_{it} at higher frequencies are due to the trap charges that are capable to follow low frequencies and as the frequency increases trap charges cannot follow it, therefore, it does not contribute to D_{it} at the larger frequencies. In all the cases SiO_2 material shows that it has the best interface quality (interface charge densities will increase with the dielectric constant value of material) with silicon substrate and Ta_2O_5 has the poorest.

TABLE 6.3 Comparative Results for 1 MHz Frequency.

Materials	D_{it} (Model)	D_{it} (TCAD Simulation)
SiO_2	5.26×10^{12}	5.26×10^{12}
Ta_2O_5	8.55×10^{12}	8.55×10^{12}

6.7 SUMMARY

An oxide thickness-dependent interfacial distribution of oxide charges in MOS device structures has been analytically studied. The flat band voltage of entirely the samples specified negative value owing to the presence of deep donor type surface states and positive interface charges, and these belongings are extremely accountable for variation in interface charge density with oxide thickness. The interface trap charge has been analyzed and measured based on standard capacitance-voltage method (general method). A sharp increment in the variation of interface charge as a function of decreasing oxide thickness is observed at the Si-SiO_2 interface and similar increase is also found by replacing SiO_2 with high-k dielectric materials. Theoretical estimate values are compared with the results obtained by the ATLAS device simulation, excellent agreements between the both are perceived. The interface charge densities also have been analyzed and measured based on conductance method. Linear increase in interface charge density has been observed as the dielectric constant of the oxide increases and comparable increase is also originated by decreasing the oxide thickness.

KEYWORDS

- **conductance method**
- **defects**
- **high-k materials**
- **C-V method**
- **interface charge densities**
- **tantalum pentoxide**
- **TCAD**

REFERENCES

Arora, N. *MOSFET Modeling for VLSI Simulation (Theory and Practice)*, World Scientific Publishers: Singapore, 2007.

Cheng, C., Horng, J.; Tsai, H. Electrical and Physical Characteristics of HfLaON-Gated Metal-Oxide-Semiconductor Capacitors with Various Nitrogen Concentration Profiles. *Microelectron. Eng.* **2011,** *88*, 159–165.

Deal, B. E.The Current Understanding of Charges in the Thermally Oxidized Silicon Structure. *J. Electrochem. Soc.* **1974,** *121*, 198–205.

Deal, B. E. Standardized Terminology for Oxide Charge Associated with Thermally Oxidized Silicon. *IEEE Trans.Electron Dev.* **1980,** *27*, 606–608.

Georgievska, L.; Novkovski, N.; Atanassova, E. C-V Analysis at Variable Frequency of MOS Structures with Different Gates Containing Hf-doped Ta_2O_5. *PhysicaMacedonica* **2012,** *61*, 13–20.

Goetzberger, A., Sze, S. M. Metal-Insulator-Semiconductor (MIS) Physics. In *Applied Solid State Science*, R. Wolfe, Ed. Academic Press: New York, 1969.

Houssa, M.; Pantisano, L.; Ragnarsson, L.; Degraeve, R.; Schram, T.; Pourtois, G.; De Gendt, S.; Groeseneken, G.; Heyns, M. M. Electrical Properties of High-k Gate Dielectrics: Challenges, Current Issues and Possible Solutions. *Mater. Sci.Eng.: R* **2006,** *51*, 37–85.

Hurley, P. K.; O'Connor, E.; Djara, V.; Monaghan, S.; Povey, I. M.; Long, R. D.; Sheehan, B.; Lin, J.; McIntyre, P. C.; Brennan, B.; Wallace, R. M.; Pemble, M. E.; Cherkaoui, K. The Characterization and Passivation of Fixed Oxide Charges and Interface States in the Al_2O_3/InGaAs MOS. *IEEE Trans. Dev. Mater. Reli.* **2013,** *13*, 429–443.

Kundu, S.; Shripathi, T.; Banerji, P. Interface Engineering with an MOCVD Grown ZnO Interface Passivation Layer for ZrO_2–GaAs Metal–Oxide–Semiconductor Devices. *Solid State Commun.* **2011,** *151*, 1881–1884.

Li, C.; Leung, C.; Lai, P.; Xu, J. Effects of Fluorine Incorporation on the Properties of Ge p-MOS Capacitors with HfTiON Dielectric. *Solid-State Electro.* **2010,** *54*, 675–679.

Lin, H.; Brammertz, G.; Martens, K.; Valicourt, G.; Negre, L.; Wang, W.; Tsai, W.; Meuris, M.; Heyns, M. The Fermi-Level Efficiency Method and Its Applications on High Interface Trap Density Oxide-Semiconductor Interfaces. *Appl. Phy. Lett.* **2009,** *94* (153508), 1–3.

Lu, C.;, Liao, K.; Lu, C.; Chang, S.; Wang, T.; Hou, C.; Hsu, Y. Tunneling Component Suppression in Charge Pumping Measurement and Reliability Study for High-k Gated MOSFETs. *Microelectron. Reliab.* **2011,** *51*, 2110–2114.

Maity, N. P.; Maity, R.; Thapa, R. K.; Baishya, S. Study of Interface Charge Densities for ZrO_2 and HfO_2 Based Metal-Oxide-Semiconductor Devices. *Adv. Mater. Sci. Eng.* **2014,** *2014*, 1–6.

Maity, N. P.; Thakur, R. R.; Maity, R.; Thapa, R. K.; Baishya, S. Analysis of Interface Charge Densities for High-k Dielectric Materials Based Metal Oxide Semiconductor Devices. *Inter. J. Nanosci.* **2016,** *15* (1660011), 1–6.

Maity, N. P.; Maity, R.; Baishya, S. An Analytical Model for the Surface Potential and Threshold Voltage of a Double-Gate Heterojunction Tunnel FinFET, *J.Comput. Electron.* **2018,** Online published (doi.org/10.1007/s10825-018-01279-5).

Maity, N. P.; Maity, R.; Maity, S.; Baishya, S. Comparative Analysis of the Quantum FinFET and Trigate FinFET Based on Modeling and Simulation, *J. Comput. Electro.* **2019,** Online published (doi.org/10.1007/s10825-018-01294-z).

Mamatrishat, M.; Kubota, T.; Seki, T.; Kakushima, K.; Ahmet, P.; Tsutsui, K.; Kataoka, Y.; Nishiyama, A.; Sugii, N.; Natori, K.; Hattori, T.; Iwai, H. Oxide and Interface Trap Densities Estimation in Ultrathin W/La_2O_3/Si MOS Capacitors. *Microelectron. Reliab.* **2012,** *52*, 1039–1042.

Martens, K.; Chui, C.; Brammertz, G.; De Jaeger, B.; Kuzum, D.; Meuris, M.; Heyns, M. M.; Krishnamohan, T.; Saraswat, K.; Maes, H. E.; Groeseneken, G. On the

Correct Extraction of Interface Charge Density of MOS Devices with High-Mobility Semiconductor Surface. *IEEE Trans. Electron Dev.* **2008,** *55*, 547–556.

Nicollian, E. H.; Brews, J. R. *MOS (Metal Oxide Semiconductor) Physics and Technology*, Wiley: New York, 2002.

Rathee, D.; Kumar, M.; Arya, S. K. CMOS Development and Optimization, Scaling Issue and Replacement with High-k Material for Future Microelectronics. *Inter. J. Compu. Appl.* **2010,** *8*, 10–17.

Schroder, D. K. *Semiconductor Material and Device Characterization*, Wiley: New York, 1990.

Shanfield, Z.; Brown, G.; Revesz, A.; Hughes, H. A New MOS Radiation-Induced Charge: Negative Fixed Interface Charge. *IEEE Trans. Nucl. Sci.* **2002,** *39*, 303–307.

Srinivasan, V.; Pandya, A. Dosimetry Aspects of Hafnium Oxide Metal-Oxide-Semiconductor (MOS) Capacitor. *Thin Solid Films* **2011,** *520*, 574–577.

Tapajna, M.; Huseková, K.; Espinos, J. P.; Harmatha, L.; Fröhlich, K. Precise Determination of Metal Effective Work Function and Fixed Oxide Charge in MOS Capacitors with High Dielectric. *Mater. Sci. Semicond. Process.* **2006,** *9*, 969–974.

Uren, M. J.; Brunson, K. M.; Hodge, A. M. Separation of Two Distinct Fast Interface State Contributions at the Si/SiO_2 Interface Using the Conductance Technique. *Appl. Phy. Lett.* **1992,** *60*, 624–625.

Wilk, G.; Wallace, R.; Anthony, J. High-k Dielectrics: Current Status and Materials Properties Considerations. *J. Appl. Phy.* **2001,** *89*, 5243–5275.

CHAPTER 7

High-k Material Processing in CMOS VLSI Technology

PARTHA PRATIM SAHU

Department of Electronics and Communication Engineering, Tezpur University (A Central University), Tezpur 784028, India
E-mail: pps@tezu.ernet.in

ABSTRACT

This chapter describes the material processing technology in CMOS VLSI technology using high dielectric constant material. Different steps in high-k processing have been discussed, which include: wafer polishing and cleaning, oxidation, diffusion, implantation, annealing, chemical vapor deposition technique, flame hydrolysis deposition technique, sputtering, lithography, metallization, etching, assembling, and packaging.

7.1 INTRODUCTION

High-*k* material processing is one of the important steps for making insulating layer and high-*k* gate layer in CMOS VLSI design (Sahu, 2013). The processing of high-*k* material includes silicon wafer preparation, wafer cleaning, oxidation of silicon, deposition of high-*k* thin film layer, lithography, and etching (Madou, 1997; George, 1992; Kawachi et al., 1983; Kim and Shin, 2002; William et al., 2003; Lee et al., 2002; Pandhumsopom et al., 1996; Vasile et al., 1994; Marxer et al., 1997; Chem et al., 2002). In this chapter, I have tried to discuss about these basic steps.

7.2 Si-WAFER POLISHING AND CLEANING

VLSI technology uses mainly silicon wafer. It requires to know silicon wafer etching, polishing, and cleaning. Chemical etching of wafer is mainly of two types—acid etching (Sahu, 2013) and alkali etching (Sahu, 2013). In the mechanical shaping operations, the surface and the edge of the wafer is damaged and contaminated. This damage and contamination can be removed by etching chemically the wafer as discussed below.

7.2.1 *Wafer Etching*

In acid etching, the wafers are introduced into an acid sink which consists of a tank to hold the etching solution. Wafer rinsing also lakes place in the sink. The etchants used is a mixture of nitric acid, hydrofluoric, and acetic acid in the ratio of 4:1:3. Etching is an oxidation reduction process. The H^+, F^-, and NO_3^- ions are the primary reacting species. The following reaction takes place:

$$3\,Si + 4\,HNO_3 + 18\,HF \rightarrow 3\,H_2SiF_6 + 4\,NO + 8\,H_2O\,.$$

The limitation of nitric acid is that it oxidizes the surface of the wafer to form silicon dioxide. In high HF solutions, the reaction is limited by the oxidation step. Etching is anisotropic and the reaction is sensitive to doping, orientation, and defect structure of the crystal. Also all oxides formed are removed by HF. In high HNO_3, solution isotropic etching takes place. The rate of reaction is diffusion limited. That is transport of reactant products to the wafer surface across a stagnant boundary layer is the controlling mechanism. Moreover, in acid etching, the dimensional uniformity is not maintained for larger wafers so alkali etching is preferred for larger wafers.

- In alkali etching, a mixture of $NaOH/H_2O$ or KOH/H_2O is used. Alkali etching is limited by reaction rate and is intrinsically anisotropic and thus orientation dependent. Excellent uniformity in the wafer surface can be achieved by alkali etching.

7.2.2 Polishing

The last step in wafer preparations is mechanical and chemical polishing of the wafer surface to obtain a smooth, surface on which integrated circuits can be made (Sahu, 2013). A schematic of polishing process is shown in Figure 7.1. Wafer polishing can be done on a single wafer in a batch processing manner. The polishing pad is made up of artificial fabric. The wafers are mounted on a fixture, processed against the pad under high pressure and are rotated. A mixture of polishing slurry and wafer, dripped on to the pad does the polishing. The slurry consists of colloidal suspension of fine SiO_2 particles in an aqueous solution of sodium hydroxide. Sodium hydroxide oxidizes the silicon and the silica partials in the slurry abrade the oxidized silicon away. The polishing rate depends on the properties of the pad, applied pressure, composition of slurry, and the rotation speed.

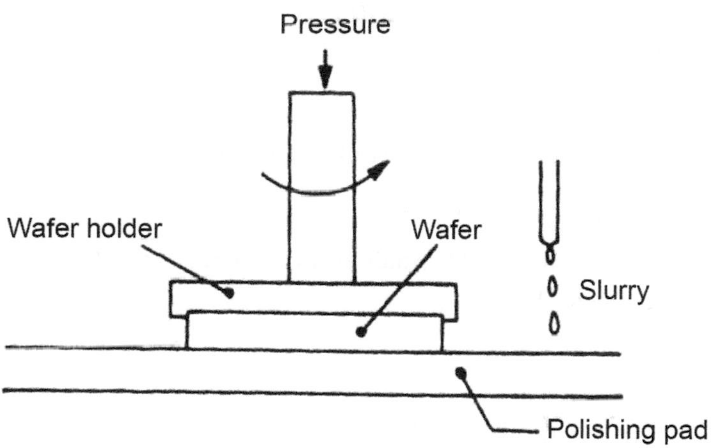

FIGURE 7.1 Wafer polishing.

7.2.3 Cleaning

The cleaning of wafer plays an important role in the fabrication process (Sahu, 2013). A thorough cleaning is necessary as dust or the impurities present may degrade the adhesion of the layers, which would be deposited further on the substrate. Cleaning process involves the following steps:

Initial method:

- 2–5 min soaked in deionized water (DI H_2O) with ultrasonic agitation.
- 30 s rinse under free-flowing DI H_2O.
- Spin rinse dry for wafers; N_2 blow off dry for tools and chucks.
- For particularly troublesome grease, oil, or wax stains, the sample was soaked in 1, 1, 1-trichloroethane (TCA) for 2–5 min or trichloroethylene (TCE) with ultrasonic agitation prior to acetone cleaning.

Acetone and methanol cleaning:

- 2–5 min soaked in acetone with ultrasonic agitation.
- 2–5 min soaked in methanol with ultrasonic agitation and dry in N_2 gas.

RCA clean:

Step 1:

- Dip the wafer in DI water: H_2O_2: NH_4OH in 5:1:1 ratio for 10 min at 75°C.
- Rinse in DI water.
- Rinse in mixture of DI water and HF in 50:1 ratio (this is done in Teflon beaker).

Step 2:

- Dip the wafer in DI + H_2O_2 + HCL in 6:1:1 ratio.
- Rinse in DI water.
- Rinse again in DI + HF in 50:1 ratio.

7.3 THERMAL OXIDATION AND OXIDATION SYSTEM

Oxide film can be grown on Si wafer by using dry oxidation or wet oxidation ambient method (oxygen + water vapor) maintaining at high elevated temperature (George, 1992; Kawachi et al., 1983). The oxidation is necessary for growing oxide layer throughout the fabrication of integrated circuits. Thermally grown silicon dioxide has reduced surface state density of silicon. There are two types of oxide layer—intrinsic and extrinsic silicon dioxide layer. Intrinsic silicon dioxide layer consists of fused silicon dioxide having melting point of 1732°C and highly stable crystalline form. Extrinsic silicon dioxide layer has made by the introduction of impurities for enhancing the properties of layer. The impurities are mainly of substitution and interstitial type. Sometimes water vapor may also be considered as an impurity present in the layer.

7.3.1 Oxidation System

Thermal oxidation has been a principal technique in silicon IC technology (Kawachi et al., 1983). Schematic cross section of a resistance heated oxidation furnace is shown in Figure 7.2a. It is described below:

- A cylindrical fused quartz tube containing the wafers is held vertically in a slotted quartz boat.
- The wafers are exposed to a source of either pure dry oxygen or pure water vapor.
- The loading end of the furnace tube enters into a vertical flow hood where a filtered flow of air is maintained.
- The hood reduces dust and particulate mailer in the air surrounding the wafers and minimizes the contamination at the time of wafer loading.
- The furnace temperature is maintained between 900°C and 1200°C.
- Microprocessors are used to regulate the gas flow sequence to control the automatic insertion and removal of silicon wafers, to control temperature, etc.

The SiO_2 can be grown very easily by heating the silicon to about 1000°C in the presence of either dry oxygen or oxygen with water vapor (wet oxygen). The resulting oxide has good uniformity, clings tenaciously to the silicon surface, and can be grown to a thickness of several micrometers without cracking. The thermal oxidation of silicon was used for the fabrication of under cladding layer of optical waveguide on silicon substrates. The layer grown by this technique exhibits good optical property with the relatively low loss.

The experimental set up consists of an open tube furnace capable of providing temperature up to 1200°C with a control of ±5°C. The ambient used for oxygen is either dry oxygen or wet oxygen. To obtain wet oxygen ambient, dry oxygen is passed through a water bubbler maintained at 95°C. The gas regulators are used to control the flow rates of oxygen and nitrogen.

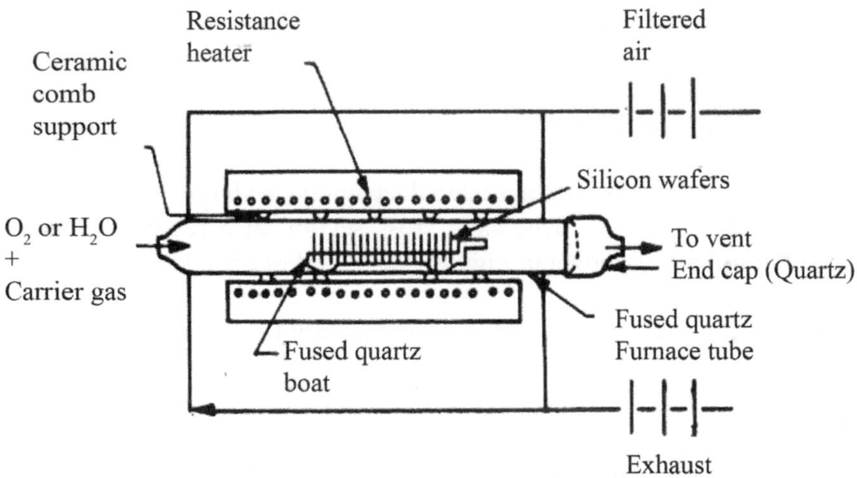

FIGURE 7.2A Experimental set up for dry or wet oxidation system.

Prior to oxidation, the silicon wafer was cleaned and finally washed in DI water. The wafers was kept in slotted quartz boat and inserted slowly to the furnace maintained at a particular desired temperature from 1000°C to 1200°C to avoid thermal stress to the wafer. The gas flow rate was adjusted within the range of 500 cc to 1 L/min depending upon the quartz tube diameter. The relation gives the oxidation time for making a sample (Sahu, 2013).

$$\frac{t}{x - x_o} = A + B(x - x_o), \tag{7.1}$$

where, t is oxidation time, x is the oxide thickness, x_0 is the oxide thickness at the beginning of oxide, and A and B are constants. The oxidation rate can be written as (Sahu, 2013):

$$\frac{dx}{dt} = \frac{1}{A + 2B(x - x_o)} \tag{7.2}$$

7.3.2 *Wet Oxidation*

For the growth of the layer, wet oxidation process was used because the growth rate in case of wet oxidation is ~3.5 times faster than that for dry oxidation (Sahu, 2013). Wet oxidation process is performed in same set up shown in Figure 7.2b, where we send wet oxygen inside furnace chamber with control manner, instead of dry oxygen. The growth rate for wet oxidation process was 0.00388 μm/min at the temperature 1100°C. Figure 7.2b shows cross section of oxidized layer formed on silicon substrate.

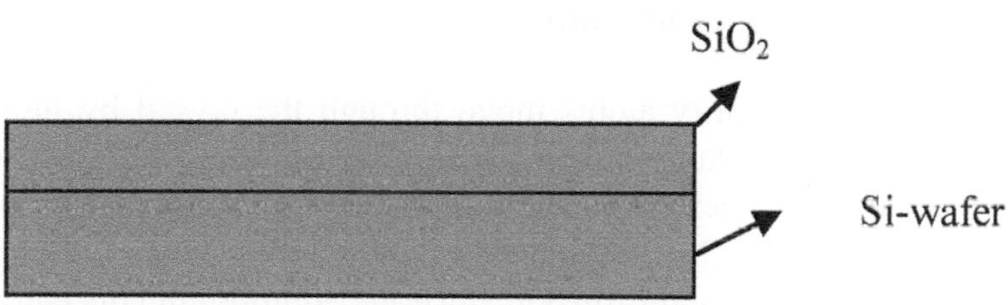

FIGURE 7.2B Cross sectional view of oxidized silicon wafer.

7.4 DIFFUSION

Diffusion is the process of introduction of controlled amount of dopant atoms into the semiconductor. Diffusion alters the type of conductivity of the semiconductor. In silicon-integrated circuit processing, diffusion is one of important step used to form base, emitter, collector, and resistors in bipolar technology and source and drain regions of MOSFET's in MOS technology (Sahu, 2013; Madou, 1997). A wide range of dopant atoms with varied concentrations can be diffused in silicon. Commonly used diffusion methods are—diffusion from a chemical source, diffusion from a doped oxide source, and diffusion and annealing from an ion implanted layer. Though precise control of dopants in terms of concentration and selectivity can be achieved by using ion implementation but still in silicon IC processing diffusion is normally used because of easy and cheaper method. In the diffusion process, the movement of atoms can be done by different ways inside the crystal lattice—interstitial diffusion, substitution diffusion, and interchange diffusion

7.4.1 Interstitial Diffusion

In this mechanism, the impurity atoms move through the crystal lattice by jumping from one interstitial site to the next as shown in Figure 7.3. This type of diffusion can also occur by a dissociative mechanism. In this case, impurity atom occupies both substitutional as well as interstitial sites. In silicon, copper, nickel, and gold are moved by this mechanism (Sahu, 2013).

7.4.2 Substitutional Diffusion

In this case, the impurity atoms move through the crystal by jumping from one lattice site to the next with substitution of the original host atom. The presence of vacancies must be required for substitutional diffusion (Sahu, 2013). This mechanism substitutional diffusion occurs at a much slower rate than interstitial diffusion since the equilibrium concentration of vacancies is low.

7.4.3 Interchange Diffusion

Two or more atoms diffuse by an interchange process to cause interchange diffusion. This is a direct process of interchange when two atoms are involved and is a cooperative interchange when larger numbers are involved. Normally, the interstitial, substitutional, and interchange mechanism can take place together in diffusion.

7.4.4 Diffusion (Impurity) Properties

Impurities of III and V groups participate in substitutional diffusion for silicon and their motion is affected by the number and the charge state of lattice point defects. These impurities are aluminum, boron, gallium, and indium of III group and antimony, arsenic, and phosphorus of V group. Impurities belonging to I and VIII group move interstitially in silicon like the alkali metals, lithium, potassium, and sodium and gases. Transition elements like cobalt, copper, gold, iron, nickel, platinum, and silver are diffused through an interstitial and substitutional mechanism. Diffusion theory is based on two main approaches (Sahu, 2013):

- The atomistic approach. It involves interactions between point defects, vacancies, interstitial atoms, and impurity atoms.
- The continuum approach. Atomic quantities are ignored and direct relations between the initial and final states can be obtained. It involves Fick's simple diffusion equation which is to be solved with proper boundary conditions and the diffusivity of the impurity atoms.

For higher impurity concentrations, diffusion profile deviates from results of simple diffusion theory (Fick's laws). To obtain the diffusion profile for higher concentrations, Flick's diffusion equation with concentration-dependent diffusivities is required. So to understand diffusion theory, we have to know the basic mechanics of diffusion along with the laws or equations that represent the diffusion of various materials in silicon.

7.4.5 Diffusion Mechanics

A pure single crystal of silicon is based on face cantered cubic lattice with a diamond structure (Sahu, 2013). Each lattice point of the crystal has a basis of two silicon atoms. The lattice has an ample amount of voids or interstitial space between the lattice atoms. However, there are always some structural imperfections or defects. The defects associated with lattice points are called point defects. Other types of defects are associated with lines, planes, or array of points. Point defect is also known as Schottky defect and is an empty lattice point in the crystal. The vacancy being created by atom that moves to the crystal surface or to an interstitial site. The combination of interstitial atom and a vacancy site thus formed is known as Frenkel defect. Larger number of Frenkel defects means smaller number of vacant sites for interstitial diffusion. With increase in temperature the point defects are also generated. If there is a concentration gradient of impurity atoms then these point defects direct atom movement or diffusion. Diffusion in solids can be understood as atomic movement of the different atoms in the crystals by vacancies or interstitials as shown in Figure 7.3.

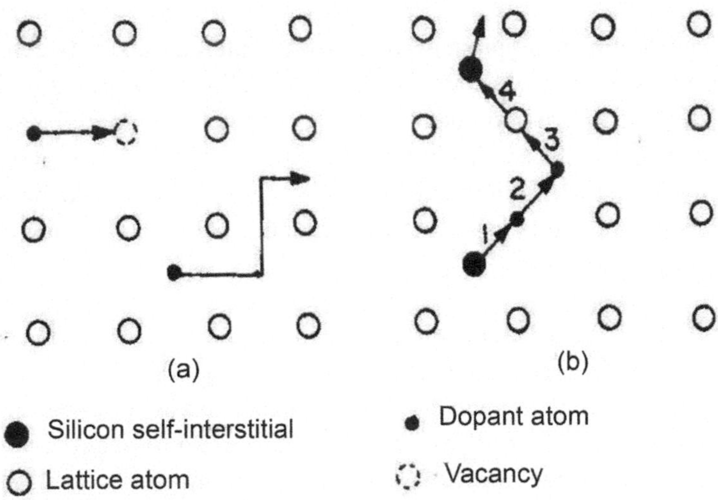

FIGURE 7.3 Different atomic diffusion mechanisms.

The lattice atoms are represented by open circles whereas the solid circles are impurity atoms. At increased temperatures the lattice atoms vibrate around their lattice sties. Some of the lattice atoms may gain sufficient energy to leave the lattice site and create a vacancy. When a neighboring atom moves to the vacancy site created by the above atoms it is known as diffusion by vacancy. If the migrating atom is a host atom, it is known as self-diffusion. If the migrating atom is an impurity atom it is known as impurity diffusion. An interstitial atom that moves from one place to another without occupying a lattice site is known as interstitial diffusion mechanism. An atom which is smaller in size than the host atoms does not form covalent bond with host and it moves interstitially. The activation energy required for the diffusion of interstitial atoms is lower than those of the host atoms by a vacancy mechanism. In the figure, a self-interstitial atom displaces an impurity atom which in term becomes an interstitial atom. The impurity atom displaces another host atom and this again becomes a self-interstitial atom. Phosphorous, boron, arsenic, and antimony diffuse in silicon through vacancy and interstitial mechanism but in boron and phosphorus interstitially component is dominating and in arsenic and antimony vacancy mechanism is dominant (Sahu, 2013; Madou, 1997). For each mechanism to operate, an imperfection or a defect is required to which the diffusing atom can move. The rate at which diffusion will take place is proportional to the number of imperfect ions present in crystal lattice.

7.4.6 *Measurement Techniques*

The result of diffusion can be obtained by conducting a few experiments to measure the junction depth and sheet resistance. Also, the measurement of diffusion profile of different dopants is important to study the way in which the dopants reside in the host lattice and in what concentration (Sahu, 2013).

7.4.6.1 *Study of Junction Depth and Sheet Resistance*

Measurement of junction depth:

The junction depth is measured on an angle lapped sample chemical staining by a mixture of 100 CC HF (49%) and a few drops of HNO_3 Al_2O_3 powder in water as the lapping mixture. This sample is put under a strong illumination for 1–2 min which would result in darker p-type region than the n region. Then with the help of interference fringe techniques one can measure the junction depth up to minimum 0.5 micron and maximum 100 microns. The result of staining is the formation of suboxide of silicon. In p-type region, since the majority carriers are the holes, the surface silicon atoms are oxidized into some form of SiO_x, ($x < 2$), that changes the surface reflectivity and gives a dark stain film of this suboxide. The stained junction is located at a place depending on the p-type concentration and the concentration gradient. A concentration in the range of mid-10^{17} atoms/cm^3 is found at the stain boundary (Lee, 2002).

Another way is to grind a cylindrical groove into the wafer is shown in Figure 7.4a. This visually magnifies the junction region which is delineated by means of a selective etch.

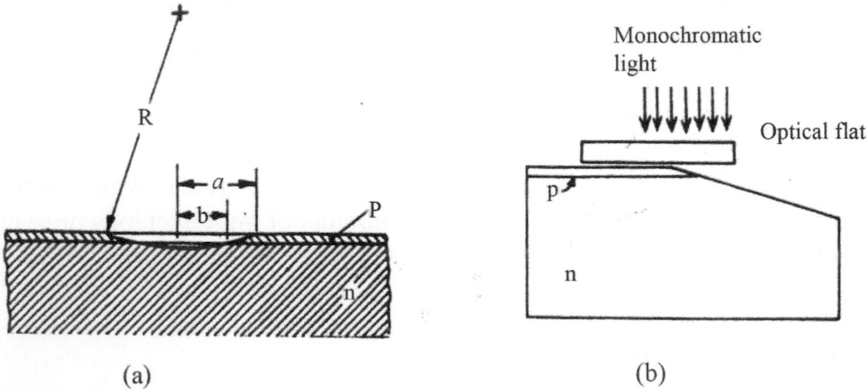

FIGURE 7.4 (a) Grinding of cylindrical groove. (b) Illumination on optical flat placed on chip.

The junction depth can be given by

$$x_j = \left(R^2 - b^2\right)^{1/2} - \left(R^2 - a^2\right)^{1/2}, \tag{7.3}$$

where *R* is the radius of the tool used for forming the groove, *a* and *b* are shown in the figure. For *R* >> *a, b* we can write:

$$x_j \approx \left(a^2 - b^2\right)/2R \tag{7.4}$$

Interferometer method is preferred with the upper surface of the chip serving as a reference plane when the lapping angle is not known accurately. An optical flat is placed on this chip and is vertically illuminated by collimated monochromatic light as shown in Figure 7.4b. The resulting fringe pattern gives a direct measure of the vertical depth in wavelengths of the illuminating source.

Sheet resistance measurement:

Figure 7.5 shows four probe methods consisting of four probe points. It is used to measure the sheet resistance of the diffused layer. The sheet resistance is expressed as (Sahu, 2013):

$$R_s = (V/I)F, \tag{7.5}$$

where V is the dc voltage across the voltage probes, I is the constant dc current passing through the current probes, and F is the correction factor. These four probe points are in line with the dc current passing through the outer two probes and the voltage is measured across the inner two probes. The average resistivity of a diffused layer is $R_s.X_j$. The correction factors F are different for different kinds of samples (circular and rectangular types). If for a circular sample with a diameter d and a rectangular sample with the side that perpendicular to the probe line is termed as 'd' and 's' as the probe spacing is taken then for large d/s (> 40), F is independent to sample types where C.F. = 4 .53. Table 7.1 shows the correction factor F for different samples.

For characterizing thin layers, sheet resistivity is measured using four point probe system (Sahu, 2013). Four probe tips are arranged in a linear array (Lee. 2002). Probe force, probe travel, tip radius, and probe material must be selected with consideration for the resistivity hardness and thickness of the layer to be measured. The outer two probes carry current and the inner probes measure the resultant voltage.

TABLE 7.1 Correction Factors for Square and Rectangular Samples.

Square				Rectangle	
d/s	**Circle**	**a/d = 1**	**a/d = 2**	**a/d = 3**	**a/d ≥ 4**
1.0				0.9988	0.9994
1.25				1.2467	1.2248
1.5			1.4788	1.4893	1.4893
1.75			1.7196	1.7238	1.7238
2.0			1.9475	1.9475	1.9475
2.5			2.3532	2.3541	2.3541
3.0	2.2662	2.4575	2.7000	2.7005	2.7005
4.0	2.9289	3.1127	3.2246	3.2248	3.2248
5.0	3.3625	3.5098	3.5749	3.5750	3.5750
7.5	3.9273	4.0095	4.0361	4.0362	4.0362
10.0	4.1716	4.2209	4.2357	4.2357	4.2357
15.0	4.3646	4.3882	4.3947	4.3947	4.3947
20.0	4.4364	4.4516	4.4553	4.4553	4.4553
40.0	4.5076	4.5120	4.5129	4.5129	4.5129
∞	4.5324	4.5324	4.5325	4.5325	4.5324

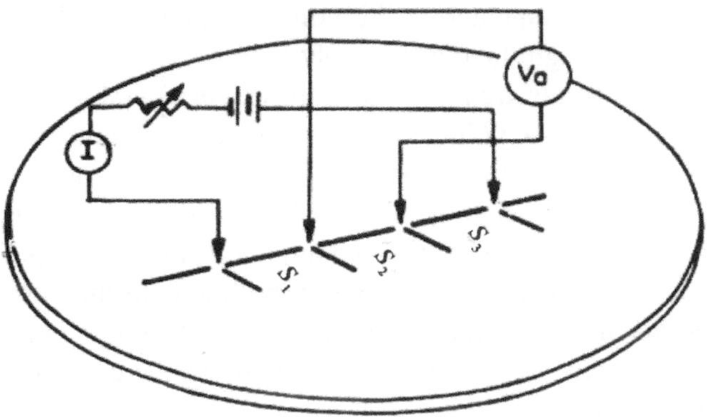

FIGURE 7.5 Four probe method.

Measurement of doping profile:

Differential conductivity technique: In this method, the diffusion profile is measured by electrical method. Repeated measurement of the sheet resistance of a diffused layer is done by four point probe technique after removing a thin layer of silicon by anodic oxidation and etching the oxide in a HF solution. In anodic oxidation at room temperature, the impurity atoms do not move and there is no segregation effect. So a doping profile can be determined.

Secondary ion mass spectrometry (SIMS) technique: SIMS technique is used in impurity diffusion profile measurement in VLSI systems and is a very accurate technique (Sahu, 2013).

Hall effect measurement: Hall effect is used for measurement of doping profile. When a piece of conductor (sample) of width w (metal and semiconductor) carrying a current, I is placed in a transverse magnetic field, an electric field E_H is produced inside the conductor in a direction normal to both the current and magnetic field as shown in Figure 7.6. This is called as Hall effect and the generated electric field is called as Hall field which gives rise to voltage is called as a Hall voltage (V_H). The Hall voltage across the sample of thickness d is obtained as (Sahu, 2013).

$$V_H = R_H . J . B . d = E_H . d \,, \qquad (7.6)$$

where R_H = Hall coefficient, J = current density = $\frac{I}{d.w}$,

$$R_H = \frac{V_H w}{I.B} \tag{7.7}$$

Measuring Hall coefficient by using the above equation and we can write, the carrier concentration,

$$n = \frac{1}{|eR_H|}, \tag{7.8}$$

where e = charge of electron.

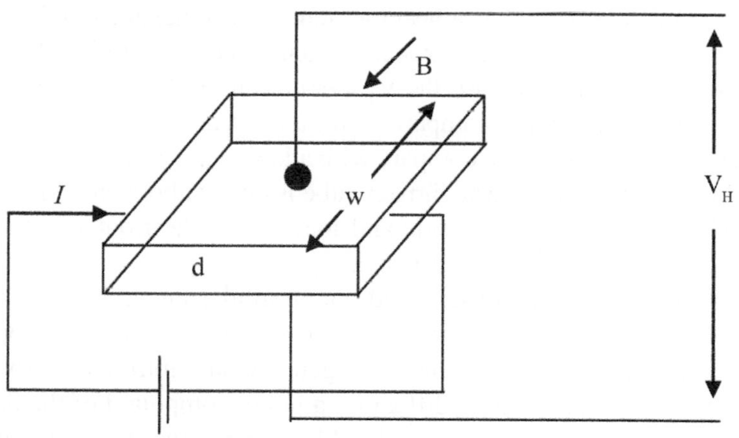

FIGURE 7.6 Set up for Hall effect measurement.

7.4.7 *Diffusion Systems*

Diffusion systems used in silicon technology are categorized as—predeposition, where the dopant is to be transported into the wafer and redistribution, where the impurity is redistributed within the wafer as a result of increased temperature. The basic requirements of any diffusion system are that a means be provided for bringing the diffusion impurity in contact with a suitably prepared wafer (Sahu, 2013). The surface concentration should be controllable over a wide range of values. Diffusion process should not damage the wafer surface. After diffusion, the material lying on the wafer surface should be easily removable. The diffusion system should give reproducible results from one diffusion run to the next and from slice to slice within a single run. The system should be capable of batch processing. There are two types of diffusion systems-sealed tube type and open tube type.

Here, the dopants and wafer are enclosed in a clean, evacuated quartz tube prior to heat treatment. After the diffusion process, the slices are removed by breaking the tube. This is a contamination-free diffusion. Sealed tube systems operate by thermal evaporation of the dopant source, transport in the gas phase, and adsorption on the surface of the semiconductor and finally the diffusion of the dopant into the wafer/slices is made. Usually, the open tube method is used where the wafer is kept in a highly temperature controlled furnace. These furnaces have a ceramic core on which Kanthal (alloy of AI, Cr, and Ni) wire is wound as a heating element. The heating coil has three separate temperature controlled zone. For temperature control, a Pt—PtRh thermocouple is used to sense the temperature in each of the zones. The central zone thermocouple compares with a reference set temperature and the outer zone thermocouple compare with additional thermocouples placed in the central zone. The quartz furnace tube is open at both ends. There are two ends—source end in rear side and load end in the front side through which wafers are loaded in a quartz boat.

The dopant is a gaseous, liquid, or a solid source. In gas source diffusion, a gaseous compound of the dopant is mixed in appropriate proportions with a carrier gas like nitrogen, etc. and introduced from the source end. In a liquid source diffusion, a liquid compound of the dopant is kept at a fixed temperature in a bubbler (near source) and the carrier gas is bubbled through it. In solid source diffusion, a solid compound of the dopant is taken. The choice of dopant source depends on the way the dopant presented to the wafer.

7.4.7.1 Solid Source Diffusion System

Figure 7.7a shows a solid source diffusion system (Sahu, 2013) where a platinum boat is used to hold a source of dopant oxide upstream from the carrier with the semiconductor wafers. The carrier gas transports vapors from this source and deposits them on the wafers. Source shut off is done by moving the dopant source to a colder region of the furnace. Sometimes two temperature furnace is also used where the source is maintained at a lower temperature than is used for diffusion. Solid sources can also be directly placed on the semiconductor slice-by chemical vapor deposition

(CVD). Deposited solid sources allow the use of very dilute dopant concentrations by adjusting the dopant to binder ratio. Diffusion from these sources results in surface concentration which is controlled by the concentration of the dopant in the oxide and not by its solid solubility limit in the semiconductor.

7.4.7.2 Liquid Source Diffusion System

Figure 7.7b shows a liquid source diffusion system (Sahu, 2013) where a carrier gas is bubbled through the liquid which is transported in vapor form to the surface of the wafer. This gas is saturated with the vapor so that the concentration is relatively independent of gas flow. Thus, the surface concentration is entirely set by the temperature of the bubbler and of the diffusion systems. Liquid source systems are convenient as compared to solid source system since the doping process can be readily initiated by control of the gas through the bubbler.

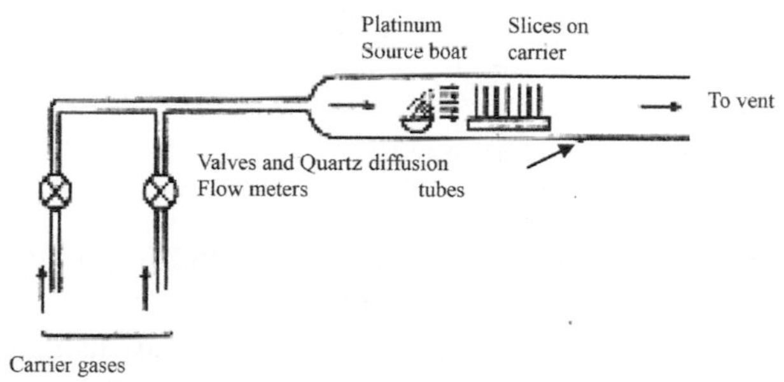

FIGURE 7.7A A solid source diffusion system.

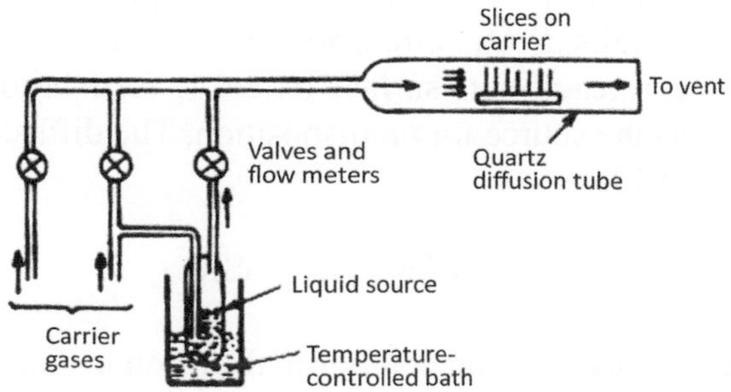

FIGURE 7.7B A liquid source diffusion system.

7.4.7.3 Gaseous Source Diffusion System

Figure 7.7c shows a gaseous source diffusion system where surface concentration can be controlled by adjusting the gas flow. Gaseous sources are more convenient than the liquid source systems. In gaseous source diffusion system provision is made for an ambient carrier gas in which the diffusion takes place. A chemical trap is often incorporated to dispose of unreacted dopant gases.

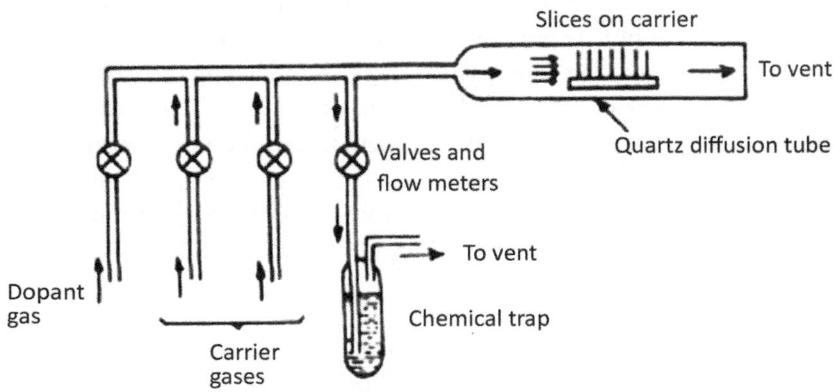

FIGURE 7.7C A solid source diffusion system.

The dopant sources for diffusion of p-type and n-type impurity and their doping mechanism are given below:

Boron: Boron is p-type impurity which is doped by using predisposition and the sources for boron doping are B_2O_3, BN, BBr_3, and B_2H_6. In open tube method for B_2O_3, the source B_2O_3 is coated on to the diffusion tube and the wafers are placed in a jig with the polished surfaces to be doped, facing the walls. B_2O_3 is transported onto the wafer as HBO_2. At the wafer surface, it reacts with surface oxides to form borosilicate glass, which later acts as the source for predisposition. The diffusion reaction is given by (Sahu, 2013):

$$2B_2O_3 + 3Si = 4B + 3SiO_2 .$$

BBr_3.is the most useful source of boron diffusion in silicon as it covers most conditions in Silicon processing. The BBr_3 transported to the wafer by the nitrogen gas which is a carrier encounters on O_2 stream which converts it to B_2O_3 in the diffusion tube. The B_2O_3 forms a boron-rich glass on the wafer surface. Here, the reaction is:

$$4BBr_3 + 3O_2 \rightarrow 2B_2O_3 + 6Br_2 .$$

The B_2H_6, the gaseous diborane is mixed with argon or nitrogen and small percentage of oxygen to convert the diborane to 82°3 which forms a boron rich glass at the surface. The preliminary oxidizing reaction is:

$$B_2H_6 + 3O_2 \rightarrow B_2O_3 + 3H_2O .$$

Higher surface concentrations can be obtained for low temperature (800°C–900°C) predeposition by using carbon dioxide rather than oxygen as (Sahu, 2013):

$$B_2H_6 + 6CO_2 \rightarrow B_2O_3 + 6CO + 3H_2O .$$

Antimony: Antimony is n type impurity which is doped by using predisposition. The diffusion is carried out by using Sb_2O_3 or elemental Sb as source. When using Sb_2O_3 as a source, the trioxide is loaded as a powder in a boat and kept in a different source furnace tube having the same axis as the main diffusion lube and kept at a temperature of 550°C–600°C. The wafers are kept downstream at a higher temperature (1250°C). Carrier gas used is $N_2$and O_2. Antimony and silicon wafers can also be placed in a vacuum inside a silica capsule which is heated to the diffusion temperature. Antimony vapors are transported to the silicon wafers and diffuse onto it.

Arsenic: Arsenic is n type impurity which is doped by using predisposition (Sahu, 2013). The diffusion sources for arsenic are arsine, arsenic trioxide, and arsenic capsules. In an open-tube gaseous source, diffusion is carried out at temperatures 1100°C–1300°C. As_2O_3 is formed by reaction of arsine with oxygen in the presence of inert carrier gas of argon or nitrogen. As_2O_3 is transported to the wafer surface where it forms an arsenic rich glass. In As_2O_3 process, the trioxide is evaporated in a source furnace and transported to the wafer by a carrier gas. The evaporation of the source takes place at around 350°C–450°C.

Phosphorus: Phosphorous is n type impurity which is doped by using predisposition. The sources for phosphorus diffusion are P_2O_5—phosphine,

$POCl_3$, and phosphorus doped silicon capsules. In $POCl_3$ systems, the source is kept in a glass bubbler at a low temperature (Sahu, 2013). Nitrogen is bubbled through and then later mixed with an N_2–O_2 mixture. $POCl_3$ reacts with O_2 to form P_2O_5 which reacts with silicon to form phosphosilicate glass. Diffusion of phosphorus takes place in a temperature range of 900°C–1200°C. It is doped in gaseous source diffusion system.

7.5 IMPLANTATION SYSTEMS

In VLSI process technology, it is required to introduce small quantities of dopant atoms. For doping in controlled manner, the doping should be reproducible and free from undesirable side effects. So to meet this needs, ion implantation method is developed. The ions are generated normally by ionization of gaseous chemical compounds and then these ion species required to be doped are selected by an appropriate magnetic fields. These ion species are then focused to form an ion beam which is accelerated by application of electric field and made to be targeted to area to be doped (Sahu, 2013).

The basic requirement for an ion implantation system is to deliver a beam of ions of a particular type and energy to the silicon surface for doping of impurities. Ion implantation machine has ion sources, accelerators, systems for ion species selection, and systems for manipulation of the ion beam over the semiconductor target or beam scanners. Since, basic ion beam requires a large vacuum system, all the subsystems are housed within a number of interconnected vacuum system as shown in Figure 7.8a. A heated filament requires the molecules to break up into charged fragments. The ion plasma contains the desired ion together with many other species. An excitation voltage accelerates the charged ions to move out of the ion source into the analyzer. The pressure is kept below 10^{-6} Torr to minimize ion scattering by gas molecules (present in tube). The magnetic field is so chosen that only ions with desired charge to mass ratio can travel through without being blocked by the analyzer walls. The ions continue to the acceleration tube for acceleration to the implantation energy as these moves from high voltage to ground. Small apertures are used to keep the beam collimated. The beam is then scanned over the surface of the wafer using electrostatic deflection plates. A commercial ion implanter is typically 6 m long, 3 m wide, and 2 m high, consumes 45 kW of power and can process 200 wafers per hour with ion dose of 1015 ion/cm/ in 100 mm wafers. A simplified block diagram of an ion implantation system is shown in Figure 7.8b. Also, a quadruple lens system is placed after the accelerator column to focus the beam on to the target (Sahu, 2013).

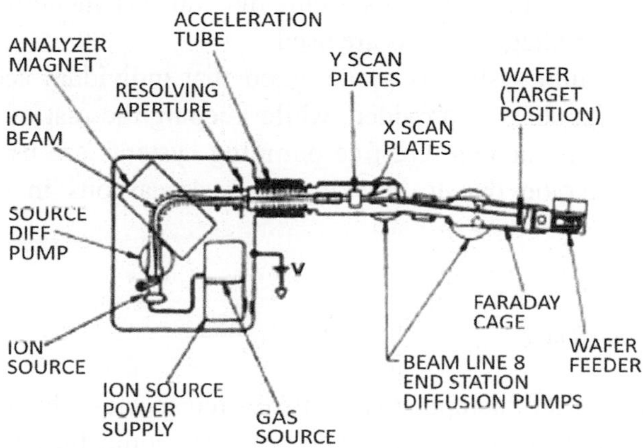

FIGURE 7.8A Ion implementation system.

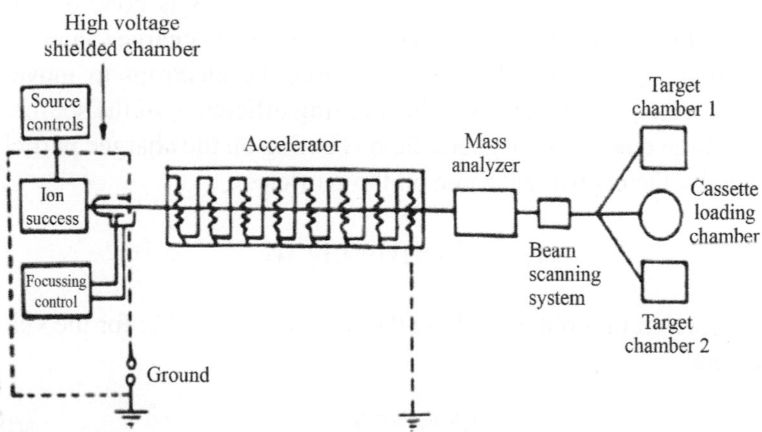

FIGURE 7.8B Schematic block of simplified ion implantation.

In order to get controlled doping to the silicon target, the following steps/systems are required:

- A wide variety of beam scanning techniques such as electrostatic, magnetic, or mechanical system are used. Also, hybrid systems consisting of all three techniques are sometimes used.
- The ion current is also monitored precisely, so that the dose of dopants introduced is also controlled.
- A variety of auxiliary subsystems like various vacuum techniques and high voltage systems are used
- The vacuum system is so designed that individual access to the target chambers is provided, while keeping the rest of the system under high vacuum. Oil free pumping systems are used to avoid contamination due to cracking of hydrocarbons in residual oil vapors.

7.5.1 Ion Sources

Ion sources consist of compounds of the desired species which are ionized before delivery to the ion accelerator column. Any ionizable compound can be used as ion source. Gaseous materials are more convenient to use than solid ones since they avoid the necessity of having to use a vaporization chamber. The dopant material must be ionized. This is done by passing the vapor through a hot or cold cathode electronic discharge (Sahu, 2013). A magnetic field is provided so as to force the electrons to move in a spiral trajectory which increases the ionizing efficiency of the source. The Lorentz force due to the magnetic field B acting on the charged particles is balanced by the centrifugal force on the particles

$$F_L = q(v \times B) = mv^2/r \tag{7.9}$$

The accelerating potential *V on* the ions is responsible for the velocity of the ions.

$$qV = ½\, mv^2 \tag{7.10}$$

So, the minimum magnetic field:

$$B_{min} = B_{min} = \frac{1}{r_{max}}\sqrt{\frac{2mV}{q}} \tag{7.11}$$

The radius of the spiralling ions *(r)* must be less than the chamber dimensions (r_{max}) the magnetic field B must exceed a minimum value to confine the ions and prevent them from striking the chamber walls. A means for extracting the ions from the discharge should also be provided which are then fed to the accelerator column. The outlet of the discharge tube is either circular or rectangular slit and it defines the cross section of the ion beam. A schematic of a Nielson source is shown in Figure 7.9.

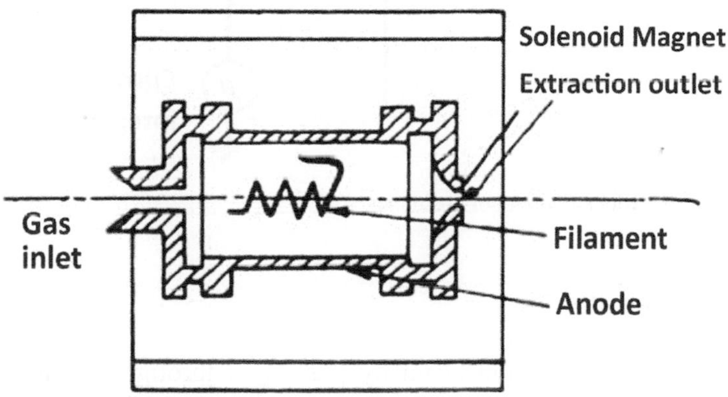

FIGURE 7.9 A schematic of a Niclson ion source.

7.5.2 *Ion Dose Measurement*

The total number of ions *N* entering the target per unit area is known as dose. If the ion beam current is *I* then, for a beam swept over a target area *A,* the dose is given by (Sahu, 2013):

$$N=\frac{1}{qA}\int I.dt \tag{7.12}$$

The circuit between target and ion source is completed through a current measuring device called dosimeter which captures all secondary electrons. A schematic of a dosimeter is given in Figure 7.10. A large current is highly desirable to allow a large number of wafers to be handled. For accurate current measurements all the secondary electrons emitted by the target by the incident ions must be recaptured.

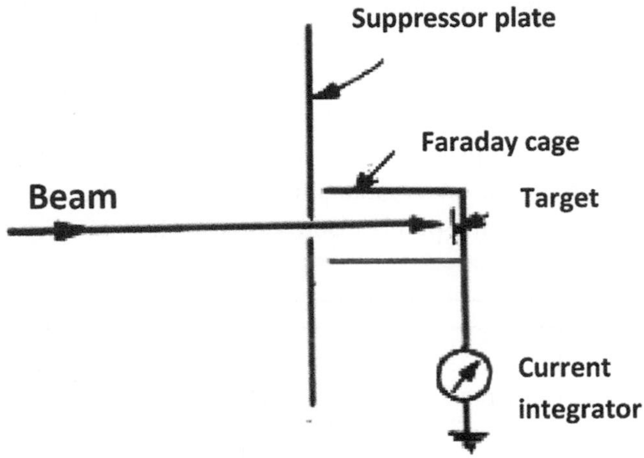

FIGURE 7.10 Block diagram of dosimeter.

7.5.3 Accelerator

Energy is given to the ion beam by passing it through a long column across by the accelerating potential. The output end of this tubular column is normally at ground potential. The energy of the beam determines the projected range of the ion. By using multiple charged ions it is possible to increase the projected range (Sahu, 2013).

7.5.4 Mass Separator

The ion source materials are usually compounds of the ion that is required for doping like BF_3, B_2H_6, and BCl_3 (Sahu, 2013), etc. By ionization, one or more charged species are generated. These ions may have other impurity contaminants. The ion separation on the basis of their masses is required for getting pure doping. By mass separation techniques, a unique distinction between ion implantation and diffusion is provided and due to this, a variety of dopants can be handled in a single machine with no effect from each other. A homogeneous field magnetic analyzer is used with the dynamics of a charged particle of mass III and velocity *v* moving at right angles to an applied magnetic field of flux density B. The particle will experience a force *F* such that:

$$F = q(v \times B) \quad (7.13)$$

This force tends to move the particle in a circular path of radius r given by:

$$r = 1/B(2mV/q)^{1/2} \quad (7.14)$$

The magnetic field B is so adjusted that radius r corresponds to the physical radius of the magnetic analyzer for the desired ion beam. So, only the desired ions will be accepted. Trajectories for different masses are shown in Figure 7.11.

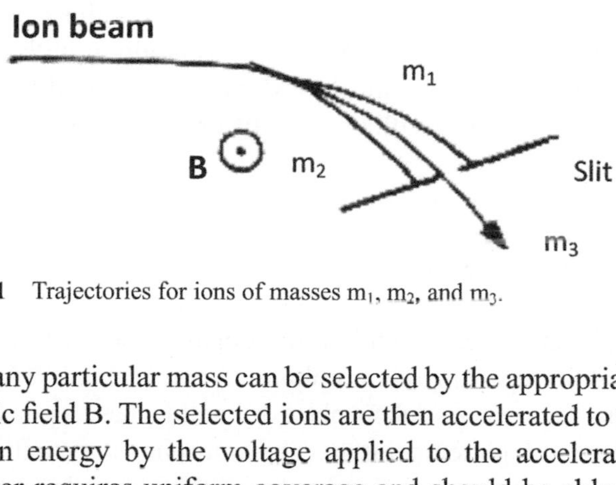

FIGURE 7.11 Trajectories for ions of masses m_1, m_2, and m_3.

Ions of any particular mass can be selected by the appropriate choice of the magnetic field B. The selected ions are then accelerated to the required implantation energy by the voltage applied to the accelerator column. An implanter requires uniform coverage and should be able to handle a number of wafers in a single pump down. Hybrid diffraction systems are generally where the beam is electronically scanned in one direction while the slices are mechanically moved in other.

7.5.5 Beam Scanning and Ion Beam Heating

The transfer of energy from the incident beam takes place which heats up the wafer. Since iIon implantation is carried out in a vacuum environment, radioactive cooling takes place. In case of radiative cooling, a temperature rise of higher order of 100°C may occur. In case of conductive cooling, the temperature rise can be limited to less than

5°C independent of the dose. So, the target wafers have a provision for conductor convective cooling. The target wafer is clamped to a sealing surface that has opening at the back for conductive cooling. The heating effects impose limitations on the ion implantation process by controlling the beam current density and hence on implantation time required to achieve a desired dose.

7.6 ANNEALING

Ion implantation damages the target and displaces many atoms from the target lattice. The electrical behavior after implantation is dominated by deep level electron and hole traps. Annealing is required to repair lattice damage and put dopant at substitutional sites where these dopant atoms will be electrically active. In VLSI, application along with repairing of damage and activate dopant annealing should minimize diffusion so that shallow implants remain shallow. The different methods for annealing are:

7.6.1 Furnace Annealing

Annealing characteristics depend on the dopant type. Furnace annealing is mainly used for repairing of amorphous silicon (Sahu, 2013). For amorphous silicon, regrowth is taken place by solid phase epitaxy which is done in furnace at high temperature. The amorphous/crystalline interface moves toward the surface at a fixed velocity that depends on temperature doping and crystal orientation. Since the activation energy for solid phase epitaxy is 2.3 eV, the process involves bond breaking at the interface. If the implantation process does not create an amorphous layer then lattice repair occurs by the generation and diffusion of point defects which requires activation energy of 5 eV and temperature of the order of 900°C to remove the defects.

Annealing an amorphous layer at high temperatures causes competition between solid phase epitaxy and local diffusive rearrangement that can lead to polysilicon formation. So, a high temperature step is preceded by a low temperature regrowth. High temperature defect diffusion can then repair extended defects remaining after solid phase epitaxy. Annealing a partially damaged layer at low temperatures can impair the process of lattice reconstruction. Since stable extended defects can be formed. When implantation does form an amorphous layer, there must be a transition region of partially damaged material at the tail of the implanted distribution. So, in this region stable defect may be formed. If the surface layer remains crystalline solid phase epitaxy can occur from both sides toward the middle and leave a string of misfit dislocations at the center due to mismatch of the layers. Incomplete annealing results in a reduction in the fraction of active dopant (Sahu, 2013). This is most apparent around the peak of the dopant distribution where damage is greatest.

7.6.2 Rapid Thermal Annealing (RTA)

RTA is used for repairing of lattice damage in single crystal silicon (Sahu, 2013). The main aim of RTA is to repair lattice damage which is a process with an activation energy of 5 eV and diffusion should be minimized which is a process involving activation energy in the range of 3–4 eV. At higher temperatures repair is faster than diffusion. RTA covers various methods of heating wafers for periods from 100 s to a few nanoseconds allowing repair of damage with minimal diffusion. RTA is done by three different ways:

Adiabatic: Here, the heating time is so short that only a thin surface film is affected. A high energy laser pulse is used to melt the surface to a depth of less than 1 μm and the surface recrystallizes by liquid phase epitaxy. Dopant diffusion in liquid state is very fast. By adjusting the pulse time and energy, shallow junctions can be obtained.

Thermal flux: This kind of annealing occurs on time scales between 10^{-7} s and 1 s, where heating from one side of the wafer with a laser, electron beam or flash lamp gives a temperature gradient across the wafer thickness. The surface is not melted but surface damage can still be repaired by solid phase epitaxy before any diffusion has time to occur.

Isothermal: This covers heating process longer than 1 s. It uses tungsten halogen lamps or graphite resistive strips to heat the wafer from one or both sides. A schematic of an isothermal rapid annealing system is shown in Figure 7.12. In this case, good activation can be obtained with less diffusion than with furnace annealing.

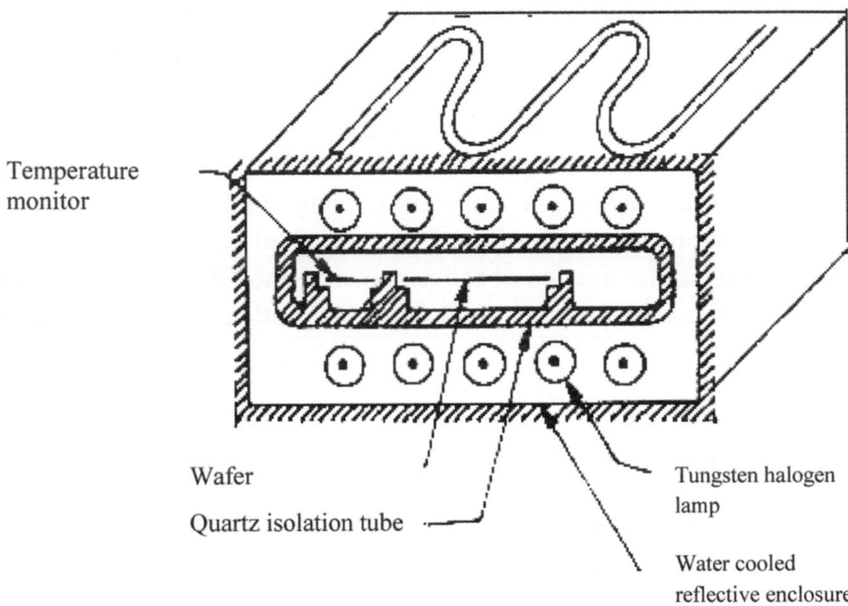

FIGURE 7.12 A schematic of an isothermal rapid annealing system.

7.7 CVD

Chemical process is used for deposition of dielectric films or removal of material from the dielectric surface (Madou, 1997). Normally, chemical deposition processes are carried out by CVD where the mass transport of the reactants are made in vapor phase and reacted and deposited on substrate. CVD technique is one of the most useful methods for the deposition of materials. In CVD, one or several gaseous species are thermally broken into their components. The components then impinge on substrate. Some of the chemical vapor components nucleate on the substrate and form thin film. The process of thermal breakdown is called as pyrolysis. The deposition of CVD is controlled by (1) mass transport and (2) reaction limited process (Madou, 1997). The various transport and reaction process are illustrated below:

1) Mass transport of reactants and diluent gases in the bulk gas flow region from the reactor inlet to the deposition zone.
2) Gas phase reactions leading to the film precursors and by-products.

3) Mass transport of film precursors and reactants to the growth surface.
4) Absorption of film precursors and reactants to the growth surface.
5) Surface migration of film formers to the growth sites.
6) Desorption of by-products of the surface reactions.

Some of the most common techniques are: metal-organic CVD (MOCVD), low pressure CVD (LPCVD), atmospheric pressure CVD (APCVD), plasma-enhanced CVD (PECVD).

7.7.1 MOCVD

This technique uses a thermally heated chamber. The sources used here are organometallic in nature. It provides thickness control within one atomic layer. Figure 7.13a shows the schematic setup for MOCVD (Sahu, 2013; Madou, 1997). The starting material (the respective acetylacetonates) in the form of very fine powder is kept in an unheated receptacle and argon gas is bubbled through it. The argon flow rate was adjusted so that the fine argon-borne particles are transported into the working chamber. Some of the fine particles settled on the substrate on holder at the center of the furnace. The temperature of the furnace is generally kept to be at ~420°C.

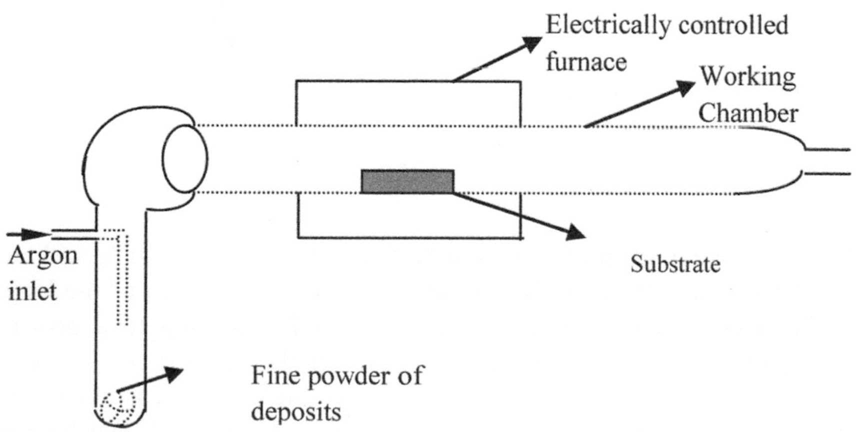

FIGURE 7.13A Schematic of MOCVD.

7.7.2 APCVD

Figure 7.13b shows schematic diagram for APCVD. This technique does not require any vacuum. At slightly reduced pressure to atmospheric (~100-10 kPa), it can be performed in continuous process with transporting a mass of deposited gas into the substrate, which is at temperature in between 300°C and 400°C. The disadvantages of the processes are poor step coverage, particle contamination (Sahu, 2013).

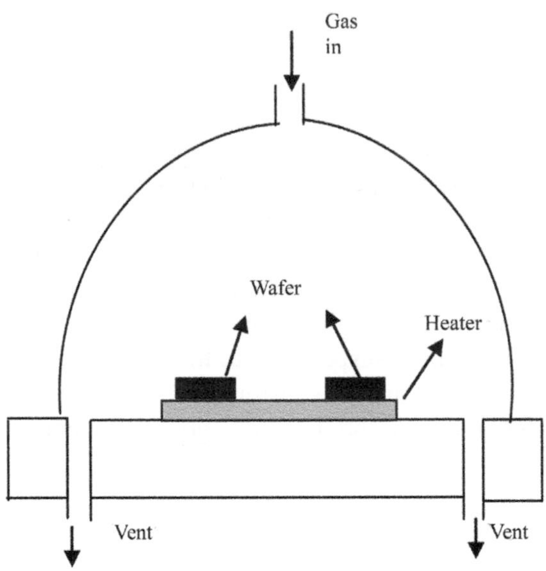

FIGURE 7.13B APCVD Machine.

7.7.3 LPCVD

It allows large number of wafers to be fabricated simultaneously without any detrimental effects to film uniformity. This is the result of the large diffusion coefficient at low pressures (~100 Pa) leading to a growth limited by the rate of surface reactions rather than by the rate of mass transfer to the substrate. The surface reaction rate is very sensitive to the temperatures but the temperature is relatively easy to control. The deposition rate of 1 μm/h have been achieved in production environment. Figure 7.13c shows the schematic diagram of LPCVD (Sahu, 2013; Madou,

1997). The substrate is placed on holder in chamber, which is heated by heater in controlled manner. The substrate is placed on holder in chamber, heated by heater in controlled manner. The substrate temperature is kept in between 500°C and 1000°C. In this case, not only the wafers/substrates but also the reaction chamber walls get coated during processing. So, it requires frequent cleaning to avoid contamination during deposition. It allows large number of wafers to be fabricated simultaneously without any detrimental effects to the film uniformity. This is the result of the large diffusion coefficient at low pressures (~100 Pa) leading to a growth limited by the rate of surface reactions rather than by the rate of mass transfer to the substrate. The surface reaction rate is very sensitive to the temperatures but the temperature is relatively easy to control. This method gives advantage of front and backside deposition. They find wide applications due to their economy, throughput and uniformity. Two main disadvantages are low deposition rate and relatively high operating temperatures.

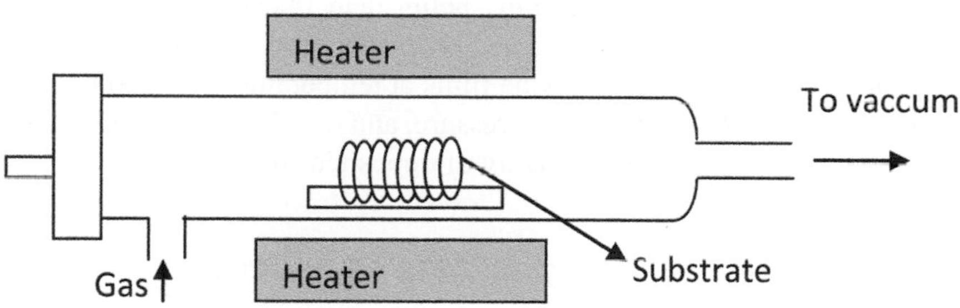

FIGURE 7.13C LPCVD machine.

7.7.3.1 Deposition of Thin Film SiON Layer by using LPCVD

The deposition of SiON layer is made from SiH_2Cl_2, O_2, and NH_3 gases with LPCVD. The processed conditions are: pressure = 100 mTorr, temperature = 900°C. The deposition rate depends on different flow rates of the reactant gases. The production of SiON film by LPCVD method with these specific parameters is mostly performed for the deposition of waveguides in higher refractive index range ($n \geq 1.7$).

$SiH_2Cl_2 + O_2 + NH_3 \rightarrow SiO_xN_yH_z$ (solid) + H_2O (gas) + HCl (gas)

SiO_xN_y (solid) + H_2O (gas) + HCl (gas)

(after annealing at temperature 1150°C), where $x = \frac{X}{X+Y}$ = concentration of oxygen, $y = \frac{X}{X+Y}$ = concentration of nitrogen, X= flow rate of O_2 and Y = flow rate of NH_3.

7.7.4 PECVD

In this method, the gases used for depositing films are dissociated by electron impact in glow discharge plasma. The substrates are placed horizontally and remain at lower temperatures than that in APCVD and LPCVD (Sahu, 2013; Madou, 1997). In this process, the RF-induced plasma ions are bombarded and transfer energy into the reactant gases. These gas particles are forced to impinge on substrate and get deposited on it due to adhesive property. This process is suitable for the deposition of doped and undoped SiO_2 and Si_2N_4. A very high deposition rate (> 1 μm/min) and refractive index uniformity better than 1% and refractive index uniformity of ~10^{-4} can be achieved with this process. This allows the deposition of relatively dense oxide films at temperature of ≤ 300°C. Low radio frequency (RF), low reactor pressure, and low deposition rate at high temperature may improve the quality of deposited film.

7.7.4.1 Deposition of SiO_2/SiON Layer by using PECVD

As stated earlier, PECVD system (Sahu, 2013) is used for the deposition of SiO_2, Si_3N_4, SiON layer due to lower temperature of operation and good uniformity of refractive index and thickness. The PECVD system used in processing of SiON layer is a parallel plate type reactor, as shown in Figure 7.13d. The glow discharge (plasma) generated by RF, takes place between two electrodes which are separated by 2 cm. The plates are of 24 cm diameter and the RF power establishing the plasma is applied to the upper electrode while the samples are placed on the bottom grounded electrode which is temperature stabilized in the range from 200°C to 350°C. The system can be operated at pressure range of 0.01–10 Torr and the applied RF power (13.56 MHz) can have values up to 300 watts. The films in the device structure are grown by making use of an intermediate state, glow discharge which contains mainly ionized gas. The ionized gas has positive, negative ions and electrons. The electron energy takes values between 1 eV and 20 eV and density varies from 10^9 to $10^{12}/cm^3$. The first step in the process of deposition is generation of reactant species via impact reactions with plasma electrons which are created by glow discharge. As for an example, electron impact reaction for SiH_4 is given below:

$$e^- + SiH_4 = SiH_2 + H_2 + e^{-1}$$

$$= SiH_3 + H_2 + e^{-1}$$

$$= SiH + 2H_2 + e^{-1}.$$

This process is observed during the deposition as a glow emitted from the plasmas. The process of deposition of solid films on a substrate using gaseous precursors involves several complicated steps. There are two extreme cases of film deposition conditions. The first case is observed at the conditions of low pressure and high RF power. In this case, the number of reactive species arriving at the surface is less than the reaction rate. This is called as mass transfer limited reaction. In second case, the number of reactive species arriving at the surface is more than the reaction rate. This is called as reaction limited process, which is observed at the low RF power and low temperature. For efficient growth of the desired film, the number of reactive species arriving at the surface should be equal to the reaction rate. For that, a careful optimization of the reaction chamber has to be conducted.

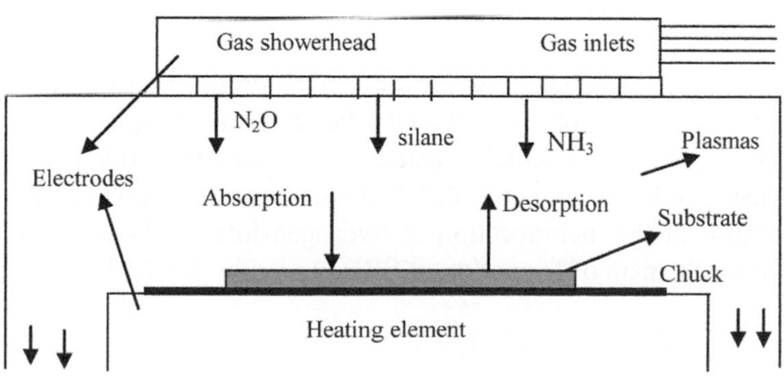

FIGURE 7.13D PECVD parallel plate plasma reactors with capacitive coupling of RF power.

7.7.4.2 Silicon dioxide (SiO_2)

The basic reaction for the formation of the silica using SiH_4 and N_2O is given below:

$$SiH_4\,(gas)+4\,N_2O=SiO_2\,(solid)+2H_20(gas)\ +4N_2\,(gas).$$

The reaction of SiH_4 with N_2O is based on the oxidation of silane by molecular oxygen produced by dissociation of N_2O. The external RF power source accelerates the electrons in the reaction chamber as shown in Figure 7.13d. The primary initial electron impact reactions between electron and reactant gases form ions and radical reactive species, that is, molecular oxygen. This molecular oxygen reacted with silane gas gives SiO_2. Thus, actually oxygen from N_2O produces SiO_2 for which refractive index becomes lower than that of SiON which is formed by N_2O and NH_3. The uniformity of the refractive index and thickness of the deposited layer are needed for small size of substrate. Silicon oxide layer was deposited on them in a single run using the following process parameters (Sahu, 2013) for a process time of 20 min.

RF power @ 13.56 MHz—10 W
Process pressure—1000 mTorr
Silane (2% SiH_4/N_2) flow rate—180 sccm
Substrate—Silicon

7.7.4.3 Silicon Nitride

Silicon nitride, other than oxide, is a material, properties of which would be of interest as we expect it to be the other extreme for silicon oxynitride layers to be grown. The stoichiometric ratio of silicon nitride is known to be 3/4 (Si_3N_4); however, again PECVD silicon nitride films deviate from this because of the incorporation of hydrogen into the layers. The basic reaction mechanism of formation of Silicon nitride (Si_3N_4) is given by:

$$SiH_4\,(gas)+4\,NH_3=Si_3N_4\,(solid\,)+7H_2\,(gas).$$

The NH_3 reacted with silane produces Si_3N_4.

7.7.4.4 SiON Layer

The optical and physical properties of SiON are expected to be between that of SiO_2 and Si_3N_4. Therefore, monitoring of the compositional characteristics of the mentioned layers should result in the better control of the SiON film properties. A general trend of decrease in the film refractive index with increasing oxygen was observed. It is because of oxygen's greater chemical reactivity compared to the nitrogen (Sahu, 2013).

From chemical reaction (2.1a) and (2.1b), it can be concluded that both N_2O and NH_3 are involved in controlling refractive index. For the deposition of SiON film, both the N_2O and NH_3 are required. The basic reaction for the formation of SiON is given below:

$$SiH_4 + N_2O + NH_3 ® SiO_xN_yH_z\,(solid) + H_2O\,(gas) + N_2\,(gas)$$

$$\rightarrow SiO_xN_y\,(solid) + H_2O\,(gas) + HCl\,(gas) \text{ (after annealing).}$$

From the above discussion, it is clear that SiON films are obtained by using silane, ammonia, and nitrous oxide as reactant gases. The different process parameters reported in for the deposition of SiON layer is shown in Table 7.2. The flow rate of silane (2% SiH_4/N_2: diluted in nitrogen because of being highly unstable at room temperature and tendency to burn in case of exposure to air) is fixed at 180 sccm and the flow rate of N_2O and NH_3 are varied to get the desired refractive index of SiON. The layer is grown at 350°C with application of RF power of 10 watts. The applied RF frequency is 13.56 MHz.

TABLE 7.2 Process Parameters of SiON Film used by Previous Author (Feridun, 2000).

Parameters	Values
Si-substrate temperature	350°C
RF power @ 13.56 MHz	10 W
Pressure	1000 mTorr
N_2O flow rate	20–450 sccm
SiH_4 flow rate	180 sccm
NH_3 flow rate	15–30 sccm

The film thickness values of the samples varied roughly between 4300°A and 3000°A. It was found that increase in N_2O flow rate results in an increase of film growth rate too. Moreover, the deposition rate was observed to be decreasing with an increase of the ammonia flow rate. These properties are attributed again to the oxygen's greater affinity for reacting with silane gas. Namely, in the first case, as the reactive oxygen concentration increases, it begins to dominate in the chemical reactions over nitrogen. As for the decrease of the growth rate with increase in nitrogen concentration in the film, the probability of the nitrogen related bonding now has increased so that nitrogen's concentration in the film increases. Thus, the layer becomes more similar to silicon nitride structured film and the growth rate is also smaller than that of silicon oxide films. In fact, if the growth rates for these films are considered, they increase in the following order: silicon nitride, silicon oxynitride, silicon oxide. So, a smooth transition of the physical properties of silicon oxynitride from those of silicon oxide to silicon nitride occurs.

The PECVD deposited SiON layer contains certain amount of O-H bonds, N-H bonds, and Si-H bonds that are known to be main cause of optical absorption at 1.38 μm, 1.48 μm, and 1.51 μm, respectively. These bonds are removed by annealing at temperature ~900°C–1000°C.

7.8 FLAME HYDROLYSIS DEPOSITION (FHD)

The origin of this process lies in the optical fiber manufacturing and can produce thick layer (~100 μm) of doped silica at high deposition rates. In addition, the deposition and consolidation process is intrinsically plane rising, hence providing excellent cladding uniformity over closely spaced cores. The process of FHD is shown in Figure 7.14. In this process, the mixture of gas is burnt in O_2/H_2 torch to produce fine particles, which stick on to substrate fixed on rotating table. $SiCl_4$, $TiCl_4$, and $GeCl_4$ are used to produce SiO_2 doped with TiO_2 or GeO_2, respectively. A small amount of Cl_2 and BCl_3 are added to lower the softening temperature of synthesized glass particles. After deposition, the heating to a temperature of around 1200°C–1300°C consolidates the material on substrate. This process has disadvantage of single side deposition only. The chemical reactions in FHD for deposition of SiO_2, $SiO_2 + GeO_2$ and $SiO_2 + TiO_2$ are given below (Kawachi et al., 1983):

SiO_2 deposition: $SiCl_4 + O_2 + 2H_2 = SiO_2 + 4HCl$
SiO_2+ GeO_2 deposition: $SiCl_4 + GeCl_4 + 2O_2 + 4H_2 = SiO_2 + GeO_2 + 8HCl$
SiO_2+TiO_2 deposition: $SiCl_4 + TiCl_4 + 2O_2 + 4H_2 = SiO_2 + TiO_2 + 8HCl$

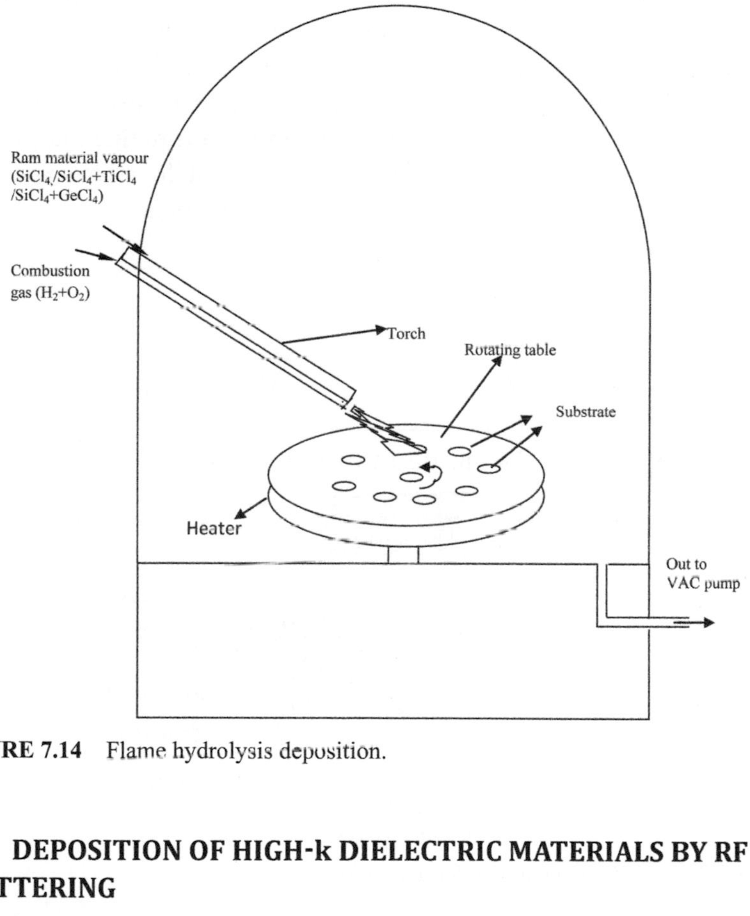

FIGURE 7.14 Flame hydrolysis deposition.

7.9 DEPOSITION OF HIGH-k DIELECTRIC MATERIALS BY RF SPUTTERING

RF Sputtering is one of commonly used method for deposition of high-*k* dielectric materials (Sahu, 2013; Madou, 1997; George, 1992; Kawachi et al., 1983). It is popular because the adhesion of the deposited material is excellent. RF sputtering is the process of ionizing the inert gas particles such Ar+ in an electric field (producing gas plasma). They are then accelerated toward the source or target through a potential gradient and the bombardment of these ions on target takes place. The energy of these gas particles physically dislodges via transfer of momentum by which atoms near surface of the target material becomes volatile and sputter off the atoms of the source material. So, sputtering is a versatile tool in which many materials can be deposited by this technique, using not only RF but also DC power. Figure 7.15 shows RF sputtering set up for deposition of high-*k* material such as oxide layers using pallets. Electrons are originated at the cathode are confined by the permanent magnets and collected by the anode. The magnetron operates at voltages with RF power in the order values less than E-beam sources. To obtain uniform film thickness appropriate mechanical motion of the substrate can be employed to expose the substrate to the same average number of sputtered atoms.

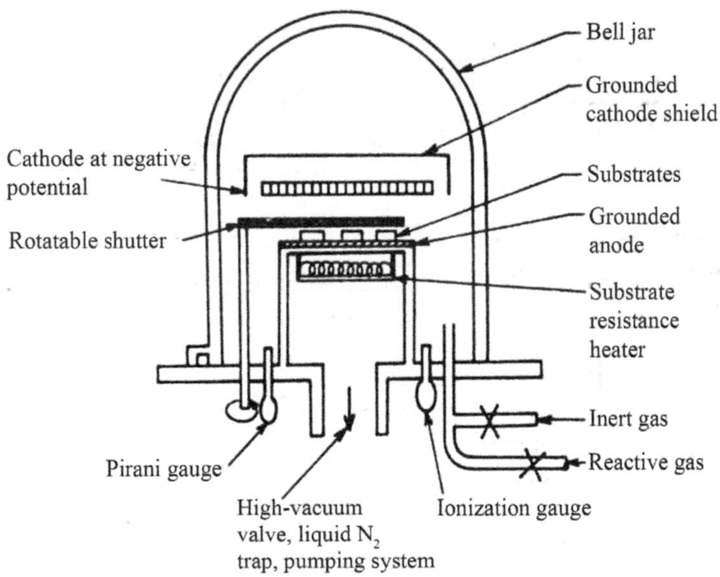

FIGURE 7.15 RF sputter system for deposition.

In this method, aluminum, silver, and gold are used for deposition. The system used for cathode sputtering is almost identical to that used for vacuum evaporation. The process however is much slower than evaporation, depositing a micron-thick film in minutes to hours compared to seconds to minutes for evaporation. However, sputtering is superior to vacuum evaporation In the quality of the film produced but sputtering equipment is more costly than most evaporation systems. The process of cathode sputtering is performed at a low pressure (about 10^{-12} Torr), the source material (material to be sputtered) is subjected to intense bombardment by the ions of a heavy inert gas such as argon.

These gas ions are usually accelerated by making the source material as the cathode of a de glow discharge. As atoms are ejected from the surface of the cathode, they diffuse away from it through a low pressure gas, depositing as a thin film on a nearby substrate. This sounds crude, but the high energy of the particles landing on the substrate actually results in a very uniform film with good crystal structure and adhesion. A potential typically 2–5 kV is applied between the cathode (source material) and anode and produces a glow discharge that fills the entire interelectrode space, except for a thin region close to the cathode.

7.10 LITHOGRAPHY

The word "lithography" is a Greek word, which literally, means stone writing. Actually, lithography is the process of transferring the pattern using optical image and photosensitive film on substrate. The basic schematic structure of lithography with alignment system is shown in Figure 7.17. There are different types of lithography electron beam lithography, ion beam lithography, X-ray lithography, and photolithography. The lithography is used to transfer the pattern with mask having high-*k* pattern developed in CMOS device. Table 7.3 shows different chemicals used as positive and negative photoresist for different lithographic process.

TABLE 7.3 Positive and Negative Photoresist for Different Lithographic Process.

Lithography process	Positive resist	Negative resist
E-beam	poly-butene-1-sulfone (PBS) poly-methyl-methacrylate (PMMA)	COP GeSe
Optical or photolithography	AZ-1350J PR102 Kodak-1813	Kodak-747 Kodak KTFR
X-ray	PBS	DCOPA

7.10.1 *Electron Beam Lithography*

Electron beam lithography provides a means to reduce image resolution below 100 nm using accelerated electrons instead of photons. Like photons, electron has also wave nature. Figure 7.16a shows the schematic diagram of electron beam lithography. When electrons are accelerated through a potential V, wavelength is given by:

$$\lambda = 1.23/ V^{1/2} \text{ where, V in KeV and } \lambda \text{ is in nm.}$$

When V = 10 KeV, λ = 0.0123nm which is much smaller than the wavelength of UV light. The electron beam lithography pattern can be transferred directly on the wafer without mask. But there are drawbacks of E-beam lithographic systems. Mainly, it needs vacuum and slow exposure speed. Collision of electrons with the substrate causes random scattering and back scattering electron generation and also secondary electrons. The scattered electrons create proximity effect in which it exposes up to several micrometers from point of impact. It is specially used for the fabrication of the mask for high resolution patterning of special devices (GaAs), SOI devices, etc. Moreover, this system is costly. Table 7.3 shows positive and negative photoresist for electron lithography (Sahu, 2013).

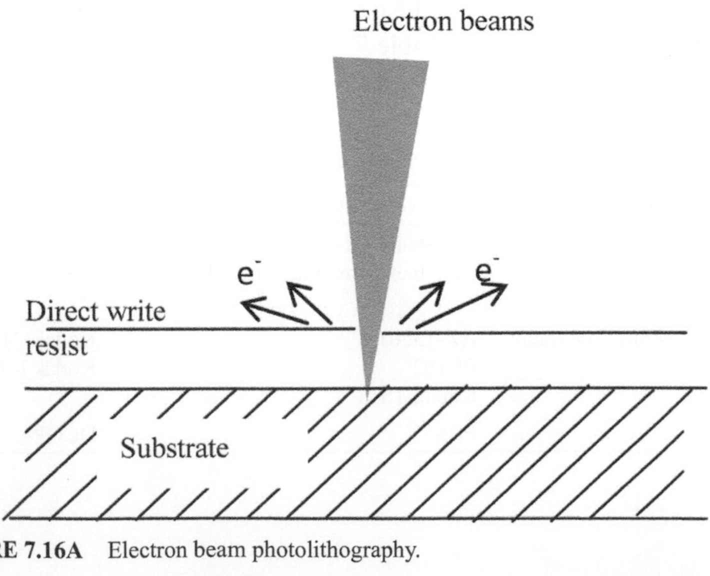

FIGURE 7.16A Electron beam photolithography.

7.10.2 X-ray Lithography

X-ray lithography is better than optical lithography because very large depth of focus (DOF) can be achieved with shorter wavelength. Figure 7.16b shows X-ray lithography process. Since, the X-ray can pass through many solids including dust and skin flakes, the clean room is necessary for chip fabrication with specifications much less than that of for UV lithography. This results in greater tolerances in dust particles. The wavelength of X-ray is order of 0.1–10 A°. The proximity mask is necessary to reduce the diffraction which increases the lifetime of mask pattern. The typical line width of 0.25–0.15 μm can be resolved using X-ray lithography with proximity mask. The main advantages of X-ray lithography are the fragility and dimensional instability of the mask and the complex alignment system. Table 7.3 shows positive and negative photoresist for X-ray lithography (Sahu, 2013).

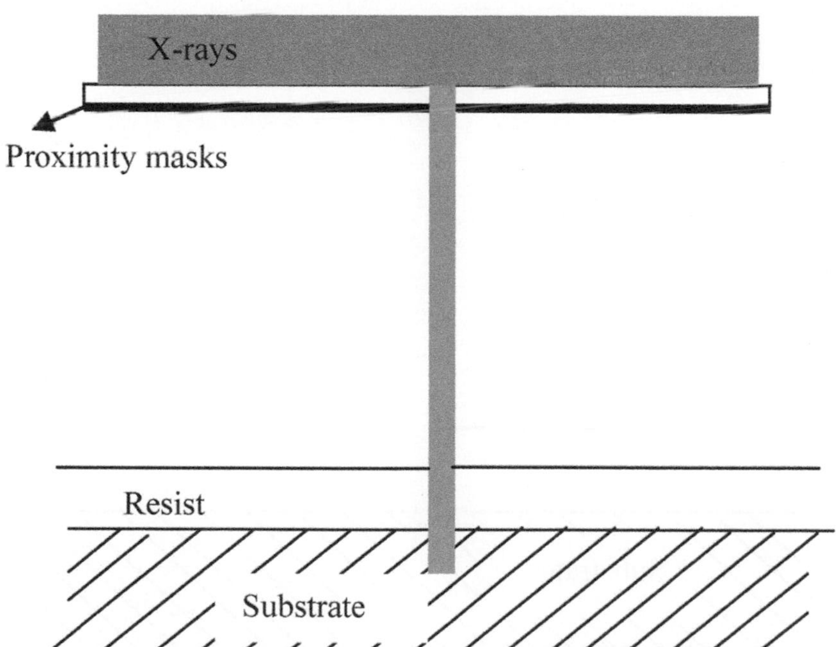

FIGURE 7.16B X-ray Lithography.

7.10.3 Ion Beam Lithography

Figure 7.16 (c) shows ion beam lithography process (Sahu, 2013). The ions beam consists of H^+, Ar^+, or He^+. These ion beams can be used to expose resists either through a mask or through broad beams by writing directly on the resist with a fine focused beam. Since, the ions are heavier than electrons, they transfer their energies more efficiently to resist and scatter much less than (<10 nm). Ion beam source also produces secondary electrons. This made high resolution with ion beam lithography without using proximity masks. The radiated beam size is 1–2 cm^2 broad. In ion beam, the diffraction effect is negligible. Resolution up to 100 nm can be achieved with it. There are two types of sources, which are used in ion beam lithography.

1) Liquid metal source
2) Gas field ionization source

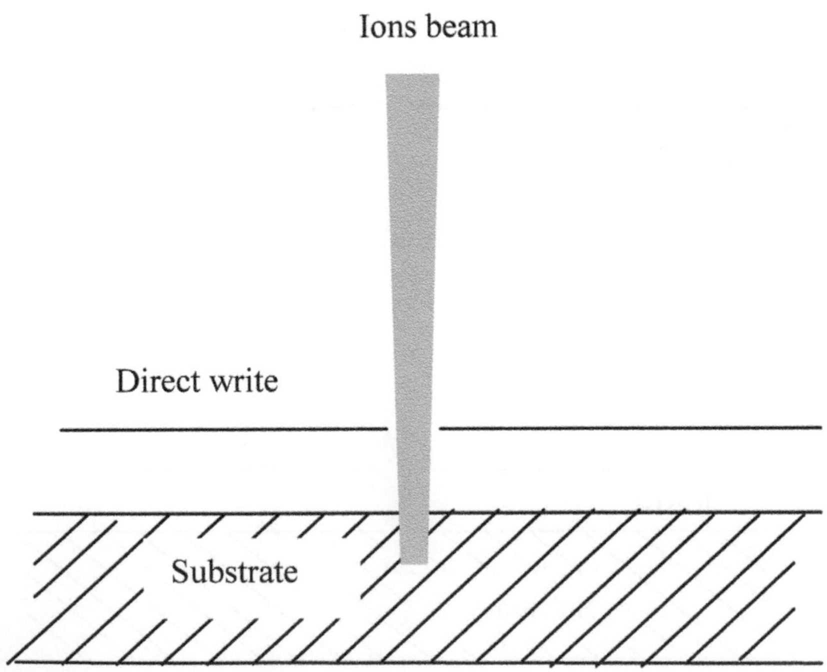

FIGURE 7.16C Ion beam lithography.

7.10.4 Photolithography

Figure 7.16d shows photolithography set up consisting of mask having device pattern, UV light, etc. In photolithography (Sahu, 2013; Madou, 1997), a film of the photoresist is first applied to the wafer. Table 7.3 shows common positive and negative photoresist used for photolithography. The radiation is shone through a transparent mask plate. The mask plate contains opaque or transparent region according to the pattern of the devices. The resulting image is focused onto the resist-coated wafer producing areas of light and shadow corresponding to the image on the mask plate. In those regions, where light was transmitted through the plate, the resist solubility is altered by photochemical reaction. In the shadowed regions, its solubility remains unaffected. This step, by analogy to photography is termed as exposure. Following exposure, the substrate is washed with a solvent that preferentially removes the resist areas of higher solubility. This step is called as development.

Depending on the type of the resist, the washed away areas may be either illuminated or the shadowed regions of the coating. If the solubility of the resist is increased with the exposure, the resist is washed away in the areas corresponding to the transparent zones of the mask plate. The resist image is identical to the opaque image on the plate and the pattern is photographic positive. Therefore resist is called as positive resist. A resist that looses the solubility when illuminated with the radiation, forms a negative images of the plates and is called as negative resist. Before going to photolithography, the following steps are followed:

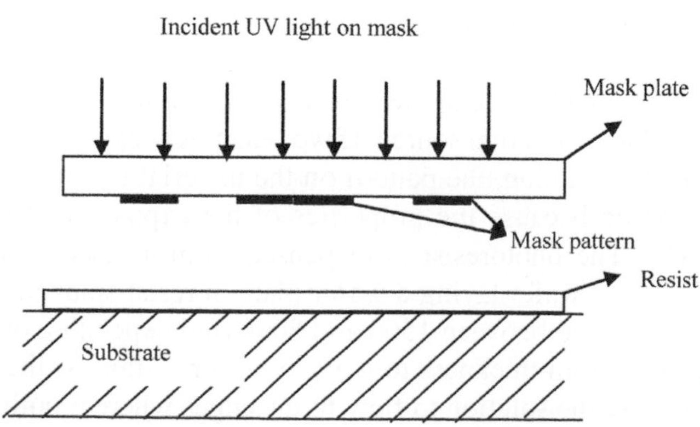

FIGURE 7.16D Photo lithography set up.

7.10.4.1 Masks Fabrication

The stencil used to generate a desired pattern in resist-coated wafers repeatedly is called as masks. In use, a photomask, a nearly optically flat glass (transparent to near UV) or quartz plate (transparent to deep UV) with a metal (e. g. 800 A° thick chromium) absorber pattern is placed into direct contact with the photoresist coated surface and the wafer is exposed to the ultraviolet radiation. The absorber pattern on the photomasks is opaque to UV radiation whereas glass/quartz is transparent.

A light field image or dark field image (shown in Figure 7.17) is then transferred to the wafer surface. This procedure results in a 1:1 image of the entire mask on to the wafer. The described masks making direct physical contact to the substrate is called as contact masks. Unfortunately, these masks degrade faster through wear than do noncontact, proximity masks (also referred as soft contact). The defects resulting from hard contact masks make it unsuitable for fabrication of device (Sahu, 2013).

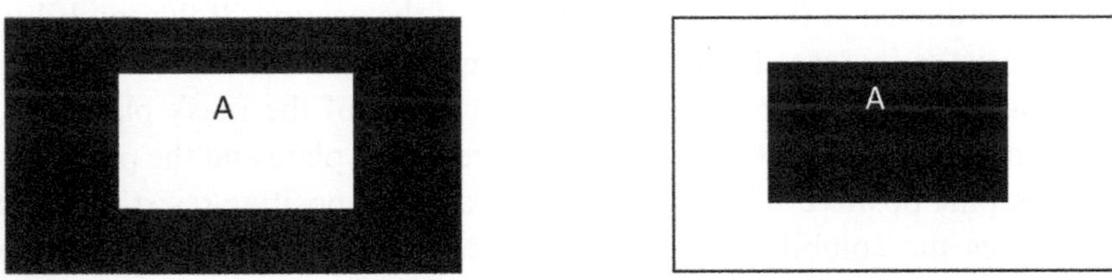

FIGURE 7.17 Types of mask: (a) dark field mask (b) bright field mask.

7.10.4.2 Spinning Resist and Soft Baking

A photosensitive material experiences a change in its physical properties when exposed to a radiation source. If we selectively expose a photosensitive material to radiation, the pattern on the material is transferred to the material exposed because the properties of the exposed and unexposed regions differ. The photoresist is dispensed form a viscous solution of polymer on to the wafer laying a wafer plate in resist spinner. The wafer is spun at high speed between 1500 and 8000 rpm depending on viscosity and the required film thickness to make a uniform film. At these speeds, centrifugal cause the solution to flow to the edges where it builds up until expelled when the surface tension is exceeded. The resulting polymer thickness (T) is a function of spin speed (w), polymer solution concentration (C in grams/100 mL solution), and viscosity (η). The empirical expression for T is given by:

$$T = \frac{K.C^{\beta}\eta^{\chi}}{w^{\alpha}}, \tag{7.15}$$

where, K = overall calibration constant. The α, β, and χ are the constants of spinning process which are determined from known thickness. After spin coating, the resist still contains up to 15% of solvent and may contain build in stresses. Figure 7.18a shows spin coating machine with operation. The coated wafers are then soft baked at 95°C for 20–25 min to remove solvent and stresses and promote adhesion of the resist layer to the wafer.

7.10.4.3 Mask Alignment

One of the important steps in photolithography process is mask alignment. The mask alignment arrangement is shown in Figure 7.19. A mask or photomask is a square glass plate with patterned emulsion of the metal film on one side. The mask is aligned with the wafer, so that the pattern can be transferred on to the wafer surface. Each mask after the first one must be aligned to the previous pattern.

7.10.4.4 Exposure

The exposure parameter is required to optimize in order to achieve accurate pattern transfer from the mask to the photosensitive layer. Figure 7.18b shows the arrangement of exposure of UV light on wafer side containing photoresist. It depends primarily on the wavelength of the radiation source. The dose of the radiation should be sufficient to get the desired change in the properties of the photoresist produced. Different photoresists exhibit different sensitivities to the different wavelengths. The dose required per unit volume of photoresist from good pattern transfer is somewhat constant. The exposure process also depends on the layer of wafer under the photoresist. For an example, a highly reflective layer under photoresist may need a higher dose than that for the lightly reflective underlying layer. The dose will also vary with the resist thickness. There are also higher order effects such as interference patterns in thick resist films on reflective wafer, which may affect the pattern quality and side wall properties.

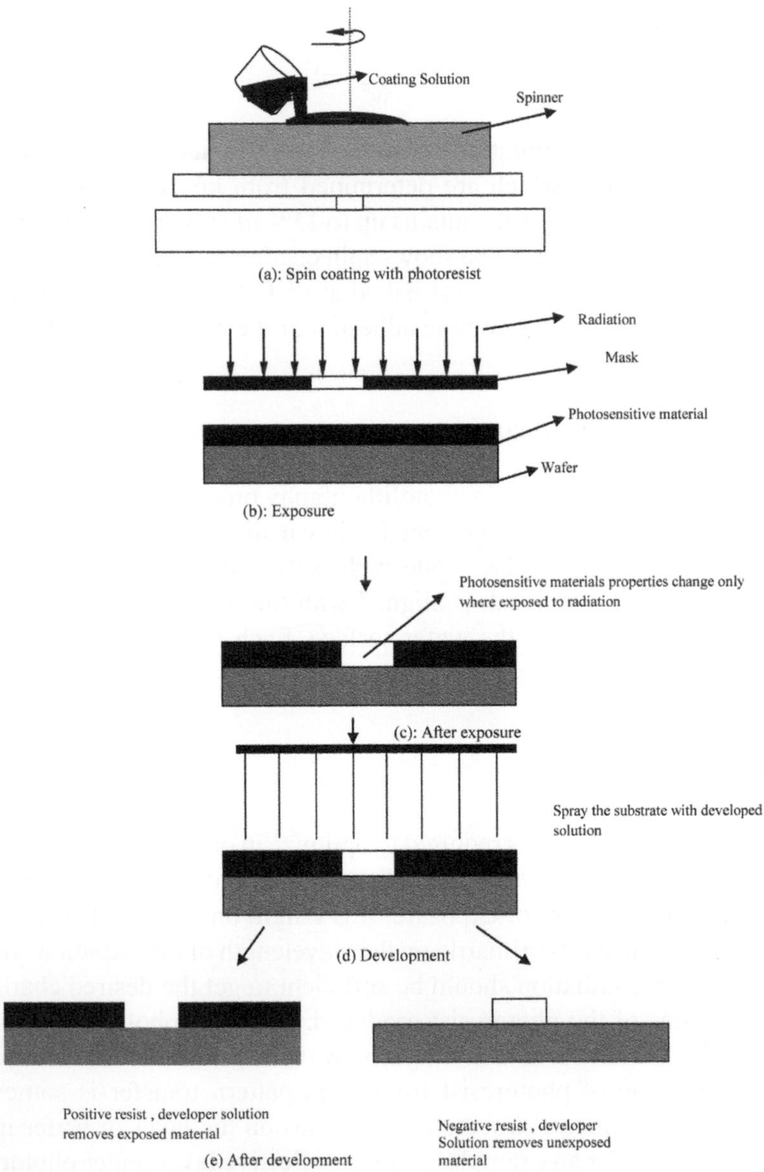

FIGURE 7.18 Steps of photolithography.

At the edges of pattern, light is scattered and diffracted. If an image is overexposed, the dose received by photoresist at the edge that should not be exposed may become significant. If we are using positive photoresist, this will result in the photoresist image being eroded along the edges, resulting in a decrease in feature size and loss of sharpness or corners, as shown in Figure 7.20. If we are using a negative resist, the photoresist image is dilated, causing the features to be larger than desired. Again accompanied by a loss of sharpness of the corners. If an image is underexposed, the pattern may not be transferred at all. Figure 7.20 shows the desired pattern, underexposed and overexposed transferred pattern on wafer.

The incident light intensity (in W/cm^2) multiplied by the exposure time (in sec) gives the incident energy (J/cm^2) across the surface of the resist film. The exposure time depend upon the quality of the mask, atmospheric condition, wafers material, etc. The optimization of the exposure time is crucial in a whole fabrication process. The quality of lithography depends on the exposure time and it requires very good accuracy especially for fabrication of low dimension waveguide core.

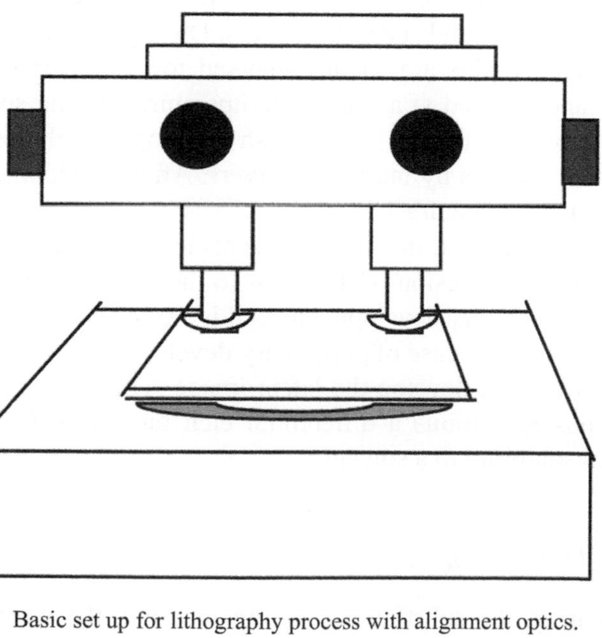

FIGURE 7.19 Basic set up for lithography process with alignment optics.

FIGURE 7.20 (a) Desired pattern, (b) overexposed, and (c) underexposed pattern.

7.10.4.5 Development

Development transforms the latent resist image formed during exposure as a relief image, which will serve as a mask for further subtractive and additive steps. During the development process, the resist dissolves selectively. Two main technologies are available for development: wet development and dry development. Wet development by solvents can be based on three different types of exposure induced changes: variation in molecular weight of the polymers (by cross linking or by chain session), reactivity change, and polarity change. Two main types of wet development setups are used: immersion and spray developers. During immersion developing, cassette-loaded wafers are batch immersed for time period in a bath of developer and agitated at a specific temperature. During spray development (as shown in Figure 7.18d, fresh developing solution is directed across wafer surfaces by fan type sprayers. After development the wafer sample is shown in Figure 7.18e.

In wet development, the use of solvents leads to some swelling of the resist and loss of adhesion of the resist to the substrate. These problems can be removed by dry development, as it is based on vapor phase process or plasma process. In case of plasma dry development oxygen ion etching (O_2-RIE) is used to develop the latent image. The latent images formed during exposure exhibits a differential etch rate to O_2-RIE rather than differential solubility to a solvent.

7.10.4.6 Post Baking

Before etching, the substrate or adding material, the wafer needs post baking. Post baking or hard baking removes residual development solvents and anneals the thin film to promote interfacial adhesion of the resist weakened by the developer penetration. Hard baking also improves the hardness of the film. The post bake frequently occurs at higher temperatures (105°C–120°C) and for longer times (30–40 min) than the prebaking step.

7.10.4.7 Ashering

A mild oxygen plasma treatment is also known as ashering that removes unwanted resist left behind after development. Resists leave thin polymer films at the resist wafer interface. The time duration of ashering depends on thickness of the resist after developing. Generally, 2–5 min is enough for ashering. It can remove resist thickness in the range of 100 A°–500 A°, depending on RF power and time for flow of 0_2.

7.11 METALLIZATION

In IC processing, metallization is one of the important steps in which proper connection of electrodes and protection of the layers is made. Aluminum (Al), Nickel (Ni), Chromium (Cr), and Nichrome (Ni-Cr) are some of the popular metals for making ohmic contact to the devices and connecting these to the bonding pads as well as a thin layer for selective etching. The adhesive properties of aluminum, nichrome are very good with silicon and silicon dioxide and they can be deposited easily (since they have low boiling point). Different metal is used for different functions related reasons. For example, nichrome is used to make the resistor in IC and also for protection layer to get selective etching. Some of the important properties of metallization process for better throughput of fabrication process are as follows:

1. Easy to form
2. Easy to etch for pattern generation
3. Stable in oxidizing ambient: oxidizable
4. Mechanically stable: good adherence, low stress
5. Surface smoothness
6. Stable throughout processing including high temperature sintering, dry or wet oxidation, etc.

7. No reaction with the final material
8. Protection from contamination
9. Increase of life time
10. Low contact resistance, low electro migration, and minimal junction penetration.

Metals layers are deposited in vacuum onto wafers by one of the following methods (Lee, 2002):

1. Filament evaporation
2. Flash evaporation
3. Electron beam evaporation
4. Sputtering.

7.11.1 Evaporation

Evaporation process can be divided in to three steps:

1. The solid metal must be changed into gaseous vapor (sublimation).
2. The gaseous metal must be transported to the substrate.
3. The gaseous metal must condense onto the substrate.

This process requires high temperature and low pressure. The chamber of the Denton vacuum must be pumped down in order to achieve the low-pressure requirement. The pressure inside the chamber is determined by the equation below:

$$P = P_0 .e^{-(S.T/V)} + Q / S, \tag{7.16}$$

where P is the chamber pressure at time t, P_0 is the initial pressure, S is the pumping speed, Q is the rate of out gassing, and V is the volume of the chamber. The concentration of gaseous metal molecules in the vacuum chamber can be determined using:

$$N = N_{av} / V = KT \tag{7.17}$$

The following quantities can be calculated for:

1. The impingement rate of the gaseous metal hitting to the substrate.

2. The time required for one monolayer of metal to deposit on the wafer surface.
3. An approximate final thickness of the metal on the substrate.

The impingement rate is given by:

$$\varphi = \frac{P}{(2\pi mkT)^{1/2}} = 2.6\times10^{20}.\,P/(MT)^{1/2}. \tag{7.18}$$

The time required for one monolayer of the metal to be deposited is given by:

$$t = N_s/\phi \tag{7.19}$$

The final thickness of the metal is given by:

$$T_h = \phi\Delta m/\rho A, \tag{7.20}$$

where ρ = density of the metal, A = area of deposition, and Δm = mass of the metal to be deposited. The evaporation process does not produce a uniform layer of Al across the substrate. The deposition rate changes as one move radically from the center of the substrate. The radial dependence of the deposition rate is described by the following equation:

$$D = D_0/\left[1+(R/H)^2\right]^{3/2} \tag{7.21}$$

7.11.1.1 Filament Evaporation

Filament evaporation is accomplished by gradually heating a filament of the metal to be evaporated. Figure 7.21a shows the schematic of filament evaporation method. This metal may come in one of the several different forms: pellets, wire, crystal, etc. Gold, Platinum, Al, Sn, Zn, Cd, Ni, and Nichrome are some mostly used for filaments evaporation to deposit on wafer. The metal is placed in molybdenum filament basket. Electrodes are connected to either side of the basket to heat and high current passed through it, causing the basket to heat. As, the heat is increased, the metallic filament is partially melts and then is eventually vaporized. In this way, atoms of the metal break free from the filament and deposit on to the wafers. It is the simplest way of metal deposition. The problem of this system is contamination during evaporation.

7.11.1.2 Flash Evaporation

Flash evaporation uses the principle of thermal-resistive heating to evaporate metals. Like filament evaporation, flash evaporation offers radiation free coatings. This technique offers some benefits over filament evaporation: contamination free coatings, speed or good throughput of wafers, and ability to coat the composite materials or layers. Flash evaporation is accomplished by passing a continuous supply of the material to be evaporated over ceramic structure that provides heat. The ceramic flash bar is heated between two positively and negatively charged posts and metal evaporates as the source material is fed on to the bar.

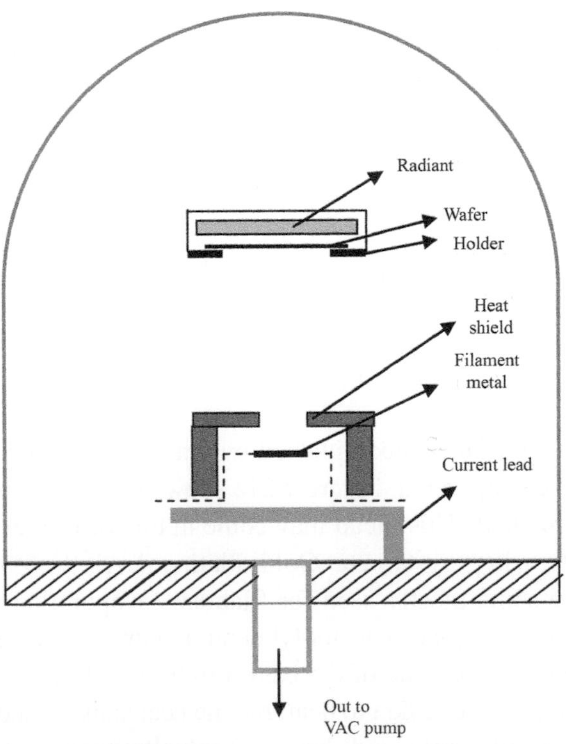

FIGURE 7.21A Filament evaporation.

7.11.1.3 Electron Beam Evaporation

Figure 7.21(b) shows the schematic of electron beam evaporation. Electron beam evaporation works by focusing an intense beam of electrons into a crucible or pocket, in the evaporator that contains Al, Ni-Cr. As the beam is directed into the source area using electric and magnetic field, the metal is heated to its melting point and eventually to evaporation temperature. The benefits of this technique are speed and low contamination, sincc thc electron beam touches the source material.

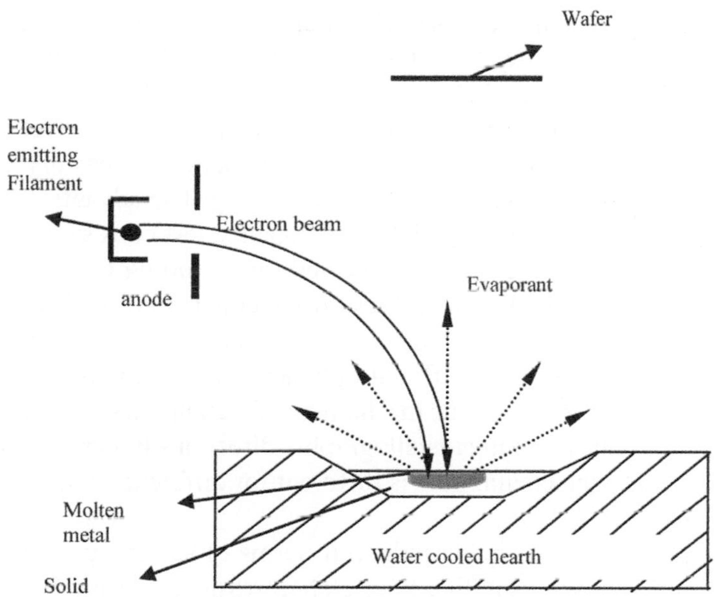

FIGURE 7.21B Electron beam deposition set up.

7.11.2 Plating Technique

The two types of plating technique used are (1) Electroplating (Sahu, 2013) (2) Electrode less plating (Sahu, 2013). Electroplating uses the typical process of coating an object with one or more layers of different metals. Cathode and anode, in this case substrate and metal, arc immersed in an electrolytic solution. When dc is passed through the solution, the positive metal ions migrate from the anode and deposit on the cathode. This method is suitable for making conduction films of gold or copper.

In electroless deposition of thin films, a metal ion in solution is reduced to the free metal and deposited as a metallic coating without the use of an electric current. This process can be used to deposit metals on any substrate such as glass, ceramic, plastic, etc., and films of considerable thickness can be deposited.

7.12 ETCHING

It is a process to remove material either selectively or nonselectively. Different methods are available for etching (Madou, 1997; William et al., 2003; Pandhumsoporn et al., 1996; Vasile et al., 1994). The most commonly used etching is dry etching and wet etching/chemical etching. Wet etching is mainly used for cleaning, shaping, and polishing whereas dry etching is used for micro machining/precision machining. The advantages of dry etching are higher aspect ratio, high directionality, less corrosion problems for metal features in the structure and less undercutting and broadening of photoresist features. The problem of dry etching is low etched rate and high sensitive to operating parameters in comparison to wet etching. So it consumes much time and is difficult to balance. Etching may be isotropic or anisotropic depending on the mechanism of etching. In case of isotropic etching, the etching rate in crystallographic directions is same whereas in anisotropy case, the etching rate is different for different crystallographic directions (Sahu, 2013).

Etch performance is being judged in terms of selectivity, uniformity, surface quality reproducibility, post etch corrosion, and profile control. Good selectivity, uniformity, and profile control are achieved at lower etch rate. In case of silica wave guide formation, the deep etching (1.5–10 μm) is required.

7.12.1 Dry Etching

Dry etching covers a family of methods by which a solid state surface is etched in the gas phase, physically by ion bombardment, chemically by chemical reaction with reactive species at the surface or combined physical and chemical mechanisms. There are different types of dry etching—plasma etching, reactive ion etching (RIE), reactive ion beam etching (RIBE), sputter etching, ion milling, barrel etching, magnetically enhanced etching, etc. Table 7.4 shows the comparison between different dry etching processes (Madou, 1997; William et al., 2003).

TABLE 7.4 Comparison Between Etching Processes.

Process	RIE	Plasma etching	RIBE	Barrel etching	Ion beam etching
Pressure in Torr	~10^{-3}–10^{-1}	~10^{-1}–10^{1}	~10^{-4}	~10^{-1}–10^{0}	~10^{-4}
Etch mechanism	Chem./Phys.	Chem.	Chem./Phys.	Chem.	Phys.
Selectivity	Good	Good	Good	Good	Poor
Profile	Anis. or Iso.	Anis./Iso.	Anis.	Iso.	Anis.

Anis., anisotropic; Iso., isotropic; Phys., physical; Chem., chemical.

7.12.1.1 Plasma Etching

Figure 7.22 shows schematic of plasma etching process (Madou, 1997). The RF frequency is generally 13.56 MHz and RF power is kept in between 1–2 kW. In plasma discharge tube, CF_4-O_2 plasmas are reacted with SiO_2 wafer. The SiO_2 are etched out through SiF_4 volatile and O_2. The pressure is kept at 10^{-1} to 1 Torr. The selectivity of etching is good because it is isotropic in nature. The chemical reaction is as follows:

$$CF_4 + SiO_2 = SiF_4 + CO_2 .$$

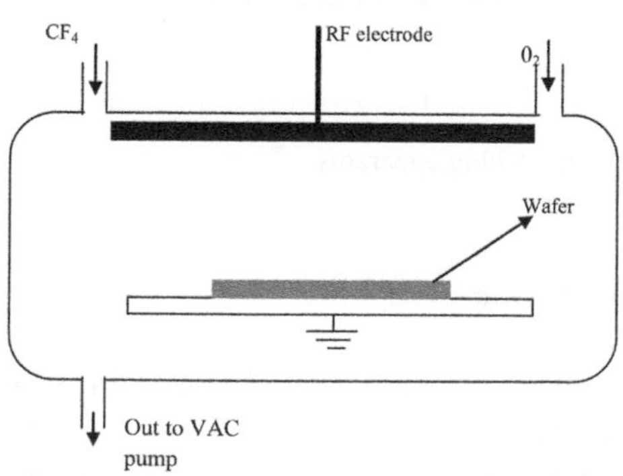

FIGURE 7.22 Plasma etching.

7.12.1.2 RIBE

Figure 7.23 shows the schematic of RIBE (William et al., 2003). It is similar to the RIE (discussed later on) except that the wafers are separated from plasmas by a grid that accelerates ions (created in the plasma) toward wafer. The ion energy is generally higher than RIE perhaps over 1 KeV which accelerates ion beam by using focusing coil. Here, only physical etch mechanism occurs for etching. The pressure is kept at 10^{-4} Torr.

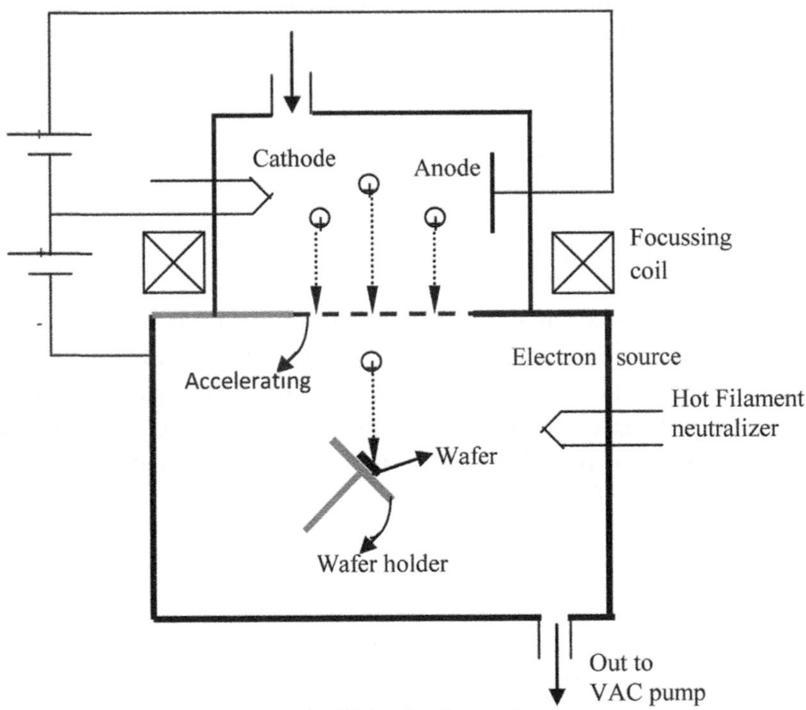

FIGURE 7.23 Ion beam etching apparatus.

7.12.1.3 Sputter Etching

It is purely mechanical phenomena. Figure 7.24 shows the schematic diagram for sputtering ion etching. In this process, energetic ions from plasma strike the substrate and physically blast the atom away from the wafer. In this process, there is no chemical reaction. Normally, the inert gases are used in this process. The inert gas ion is produces trough ionization by plasma electrons. As for example, Argon ions (Ar^{++}) are produced by simple ionization with electron impact

$$Ar + e^{-} = Ar^{++} + 2e^{-}$$

It is also called as ion beam etching.

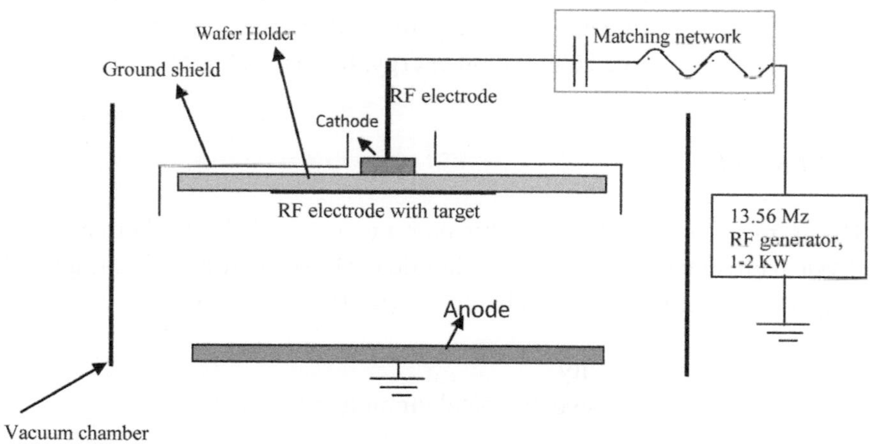

FIGURE 7.24 Sputtering etching apparatus.

7.12.1.4 Ion Milling

It is also a mechanical phenomenon (Vasile et al., 1994). However, the energetic ions are used to create the surface by bombardment. It is much more like a particle accelerator. The ions are generated, roughly collimated, and impinge on selected area of wafer placed some distance away from RF circuitry. The ions are inert gas ions. Table 7.5 shows target materials and their etching rates.

TABLE 7.5 Target of Ion Mill and their Etching Rates.

Target material	Etchant abrev.	Etchant	Etching rate
Silicon	Ion mill	Argon ions at 500 V, ~1 mA/cm^2, normal incidence	~38 nm/min
Aluminium	Ion mill	Argon ions at 500 V, ~1 mA/cm^2, normal incidence	~73 nm/min
Chromium	Ion mill	Argon ions at 500 V, ~1 mA/cm^2, normal incidence	~58 nm/min
Nichrome	Ion mill	Argon ions at 500 V, ~1 mA/cm^2, normal incidence	~18 nm/min

Ion milling is also unique in that the angle of incidence of the ions is selectable by adjusting angle between the wafer and ion beam. Adjusting the accelerated voltage controls ion energy, that is, etch rate.

7.12.1.5 RIE

RIE is most commonly used dry etching in which both chemical and physical erosion occurs. So, it is a chemical-physical process. It has gained a separate identity of its own characteristics (Sahu, 2013).

1. High directionality
2. Kinetically assisted chemical etching

Figure 7.25 shows the schematic diagram for RIE. The reactive ion species generated from plasma are used for etching. These ions have both a physical component (ion impact) and a chemical component (reactive etching). The impinging ions are bombarded on wafer and erode the surface by sputtering etch through the momentum transfer and chemical reaction. The chemical reaction takes place at low pressure of 10^{-4} to 10^{-1} Torr. Table 7.6 presents different reactive ion etchants with etching targets and rates (Pandhumsoporn et al., 1996).

Here, we used RF power which is fed into the cathode (substrate) of the parallel planar electrodes to produce reactive gas plasma at low pressure. The cathode was self-biased at the negative potential, and the electric field yielded in the vicinity of the sample surface accelerates the positive ions in the plasma toward the sample. The chemical and physical mechanisms were therefore incorporated in the etching. The RIE offers relatively large selectivity and anisotropy and is widely used for OIC fabrications. The reactant gases, $CHF_3 + O_2$ were used for SiON compound etching. Various reactive gases and additional H_2 and O_2 were used to improve selectivity and/or enhance anisotropy through suppression of the chemically reactive radicals.

The system was optimized during dummy wafer. The Ar plasmas were used for etching. Table 7.6 shows the process parameters for RIE. Thc flow rate of Ar and CHF_3 were 2.5 sccm and 5 sccm, respectively. The vacuum was order of $\sim 5 \times 10^{-5}$ Torr. The applied RF power of frequency 13.56 MHz was 215 W. The reflected RF power was 15 W with negative self bias varied between 540 V and 560 V.

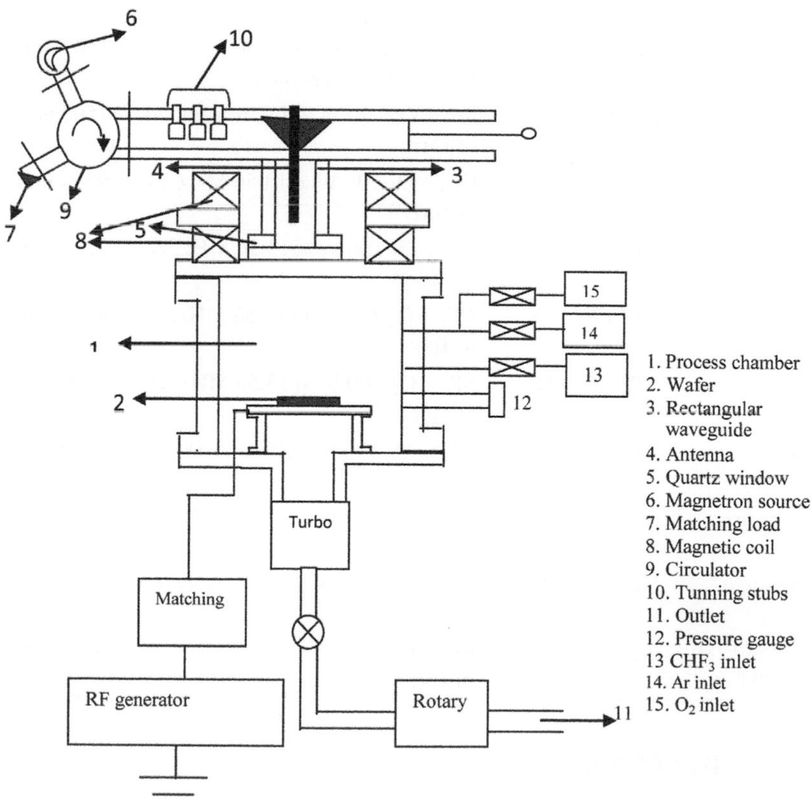

FIGURE 7.25 Reactive ion etching apparatus.

As for example, in case of Si wafers, the gases used are Cl_2/ F_2/$CFCl_3$/ CF_4+O_2, in case of SiO_2, the gases are CF_4/ CF_4+O_2/ CF_4+H_2/CHF_3. The chemical reaction for CF_4+O_2 etchant is given below (Pandhumsoporn et al., 1996):

For Si wafer with CF_4+O_2,

$CF_4 + Si + O_2 = SiF_4 + CO_2$

For SiO_2 wafer with CF_4, the chemical reaction is:

$CF_4 + SiO_2 = SiF_4 + CO_2$

TABLE 7.6 Reactive Ion Etchant and their Target Materials.

Target material	Etchant abrev.	Etchant	Etching rate
Silicon	DRIE HF Mech	Mechanical chuck, high frequency, typical recipe	~1500 nm/min
Silicon	DRIE HF ES	Electrostatic struck, high frequency, typical recipe	~1500 nm/min
Silicon	STS $SF_6 + O_2$	$SF_6 + O_2$ 100 W @13.56 MHz, 20 m Torr	~1500 nm/min
Silicon	STS $CF_4 + O_2$	$CF_4 + O_2$ 100 W @13.56 MHz, 60 m Torr	~95 nm/min
SiO_2	STS $SF_6 + O_2$	$SF_6 + O_2$ 100 W @13.56 MHz, 20 m Torr	~32 nm/min
SiO_2	STS $CF_4 + O_2$	$CF_4 + O_2$ 100 W @13.56 MHz, 60 m Torr	~43 nm/min
SiN	STS $SF_6 + O_2$	$SF_6 + O_2$ 100 W @13.56 MHz, 20 m Torr	~190 nm/min
SiN	STS $CF_4 + O_2$	$CF_4 + O_2$ 100 W @13.56 MHz, 60 m Torr	~110 nm/min
Chromium	STS $SF_6 + O_2$	$SF_6 + O_2$ 100 W @13.56 MHz, 20 m Torr	<1 nm/min
Chromium	STS $CF_4 + O_2$	$CF_4 + O_2$ 100 W @13.56 MHz, 60 m Torr	<1.3 nm/min
Nichrome (80 Ni + 20 Cr)	STS $SF_6 + O_2$	$SF_6 + O_2$ 100 W @13.56 MHz, 20 m Torr	~3.7 nm/min

7.12.2 Wet Etching

This process is called as chemical etching. In case of fabrication of planar waveguide device, it is used for metal etching. Figure 7.26 shows typical wet etching process. It proceeds by reactant transport to the surface: (1) surface reaction (2) reaction product transport away from the surface (3) stirring the solution for increase of (1) or (2). The etching rate depends on temperature, etching material, and solution composition. Diffusion limited processes have lower activation energies (the order of Kcal/mol) than reaction rate controlled processes and therefore are relatively insensitive to temperature variations. In general, one prefers reaction rate limitation, as it is easier to reproduce a temperature setting than stirring rate. The etching apparatus needs to have a good temperature controller and reliable stirring facilities. Isotropic etchants also polishing etchants, etch in all crystallographic directions at the same rate: they usually are acidic, such as $HF/HNO_3/CH_3COOH$ (HNA) and lead to rounded isotropic features in single crystalline, Si. Table 7.7 shows different wet etchants with etching targets and rates (Pandhumsoporn et al., 1996).

TABLE 7.7 Different Wet Etchants and Target Materials.

Target materiel	Etchant abrev.	Etchant chemicals and etching temperature	Etch rate
Silicon	KOH	KOH (30% by weight), ~80°C	~1100 nm/min
SiO_2	10:1 HF	10:1 HF(10 H_2O: 1 49% HF), ~20°C	~470 nm/min
Silicon nitride	Phosphoric	Phosphoric acid(85% by weight), ~160°C	~20nm/min
Aluminum	Al etch A	80% H_3PO_4 + 5% HNO_3 +5% Hac + 10% H_20, ~50°C	>500 nm/min
Chromium	CR-7	9% $(NH_4)_2Ce(NO_3)_6$ + 6% $HClO_4$ + H_2O, ~20°C	~170 nm/min
Chromium	CR-14	22% $(NH_4)_2Ce(NO_3)_6$ + 8% HAc + H_2O, ~20°C	~93 nm/min
Tungsten	H_2O_2, 50°C	30 wt % H_2O_2, 70wt% H_2O, ~50°C	~150 nm/min
Copper	Cu $FeCl_3$ 200	CE-200 (30% $FeCl_3$ + 3–4% HCl + H_20), ~20°C	~3900 nm/min
Gold	Au-5	5% I_2 + 10% KI + 85% H_2O, ~20°C	~660 nm/min
NiCr	NiCr TFN	10–20% $(NH_4)_2Ce(NO_3)_6$ + 5–6% HNO_3 + H_2O, ~20°C	~83 nm/min
Positive photoresist	Acetone	Acetone, ~20°C	~120 μm/min
Negative photoresist	Acetone	Acetone, ~20°C	~87 μm/min

In case of wet etching, the surface reaction takes place between the target material and etchant chemicals. Through diffusion the products are removed from the surface. Some of chemical reactions between target material and etchant are given below:

For SiO_2 target:

$$SiO_2 + 6HF = H_2Si\,F_6\,(aq) + 2H_2O\,.$$

For Cr target:

$$Cr + 3(NH_4)_2\,Ce(IV)(NO_3)_6 = Cr(III)(NO_3)_3\,(aq) + 3Ce(III)(NH_4)_2\,(NO_3)_5\,(aq)$$

For Cu target,

$$Cu + 2FeCl_3 + 3HCl$$
$$CuCl_3 + 2FeCl_2 + H_2$$

For Au target:

$$Au + I_2 AuI\,(aq)$$

FIGURE 7.26 Wet/chemical etching.

7.13 ASSEMBLY

Before assembly and packaging, front end of line or back end of line technique is made according to the specific requirements of a particular chip application. In front end of line, steps such as oxidation, diffusion implantation, epitaxy, lithography etching, etc. are almost same and masks used are also same as those for manufacturers whereas in back end of line these steps may be different for those of manufacturers. These designs are divided in two parts:

- Semi custom design: The masks for fabrications may not be determined by the designers.
- Full custom design: The masks for fabrications may not be determined by the designers.

After fabrication of IC, proper assembly and packaging is required for easy using, thermal stability and protection of IC in mass production point of view Figure 7.27 shows steps used in assembly for plastic or ceramic packaging.

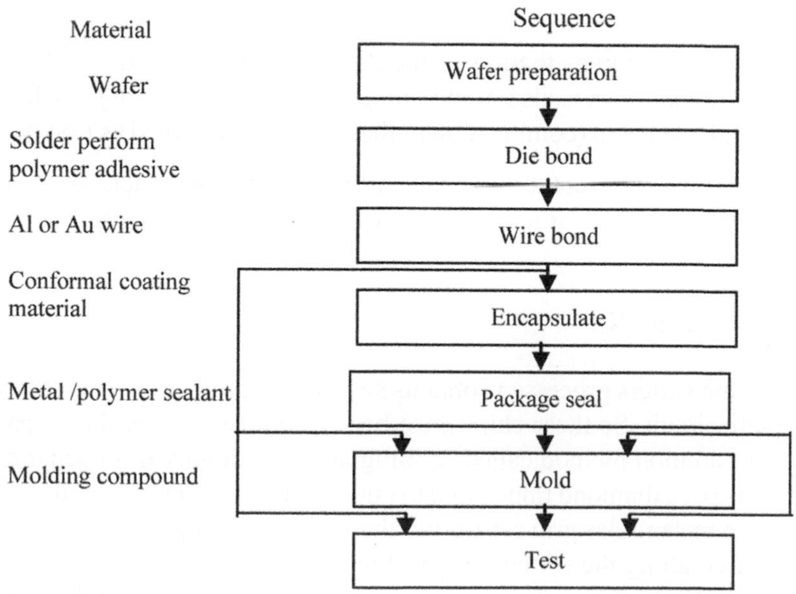

FIGURE 7.27 Steps of assembly.

Wafer preparation: Wafer backgrinding, wafer slicing, and wafer handling are done in this step (Pandhumsoporn et al., 1996).

Die bonding: Thermal stress management is made by using eutectic/hermetic ceramic and polymer or plastic technologies. In case of eutectic die bonding, the die is attached to a substrate material or to a ceramic substrate. Metallization is required on the back of the die to make it wettable by die bonding perform which is thin sheet of the appropriate solder bonding alloy. Solder die bonding to ceramic packages that hermetically sealed is performed with gold.

Wire bonding: Normally, wire bonding is made by using ball wedge bonding which utilizes the symmetrical geometry of the capillary tip. The ball bond is made in the inner portion of the tip and then wedge bond is performed around the ball bond.

Encapsulate: Surface die coating is used for encapsulation for protection from atmospheric contaminants. Silicones are normally used for surface die coating.

Package seal: Packaging seal provides the chip with physical support and environmental protection. Different types of packaging are discussed in the next section.

Molding: In molding, thermoset molding material is used for thermal stability. Epoxy resin is also used to mold ICs with condensing epichlorohydrin and bisphenol (combination of these material is called as Epoxy-A).

Test: There are different testing made after assembly and packaging. Wire bond pull test, ball shear test are some of test which may be used.

7.14 PACKAGING

Each of the wafers processed contains several hundred chips, each being a complete circuit. So these chips must be separated and individually packaged. A common method called scribing and cleaving used for separation makes use of a diamond tipped tool to cut lines into the surface of the wafer along the rectangular grid separating the individual chips. Then the wafer is fractured along the scribe lines and the individual chips are physically separated. Each chip is then mounted on a ceramic wafer and attached to a suitable package. There are three different package configurations available (Pandhumsoporn et al., 1996).

1. Glass metal package
2. Ceramic flat package
3. Dual-in-line (ceramic or plastic type).

7.14.1 Glass Metal Package

Glass and metal can package is used for simpler chip s with 5–12 leads connection Since the leads/pins are to be isolated from each other, the leads are taken out through glass metal seals. The metal header is made up of an alloy of iron, nickel, and cobalt whose thermal expansion is matched with that of glass. Figure 28a shows an example of glass metal package (Pandhumsoporn et al., 1996).

7.14.2 Ceramic Flat Package

Figure 28b shows ceramic flat package in which pins/leads are isolated with ceramic seal material and flat pins are used. It is available with 8–16 pins even with 24 or 36 or 42 pins. Because of use of ceramic material, this type of packaging is more costly than glass metal packaging.

7.14.3 Dual In-Line Package

Dual in-line packaging (DIP) is made with ceramic and plastic molded materials. It consists of three component parts: top, lead fame, and base materials. The top is hermetically sealed whereas the lead frames are made with tin plating and fixed to the base. It is also commonly available in 8, 14, or 16 leads, but even 24 or 36 or 42 leads are also available for special circuits. Most of the general purposes ICs are dual in-line plastic packages due to economy.

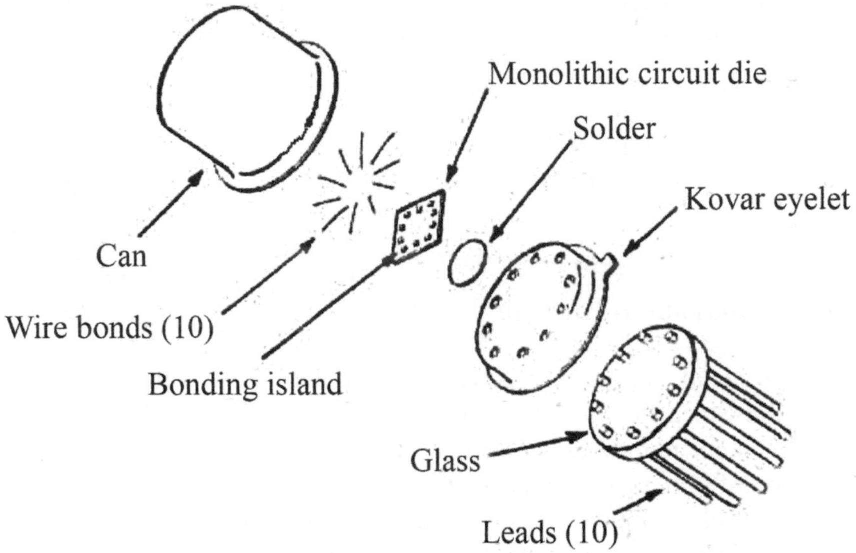

(a)

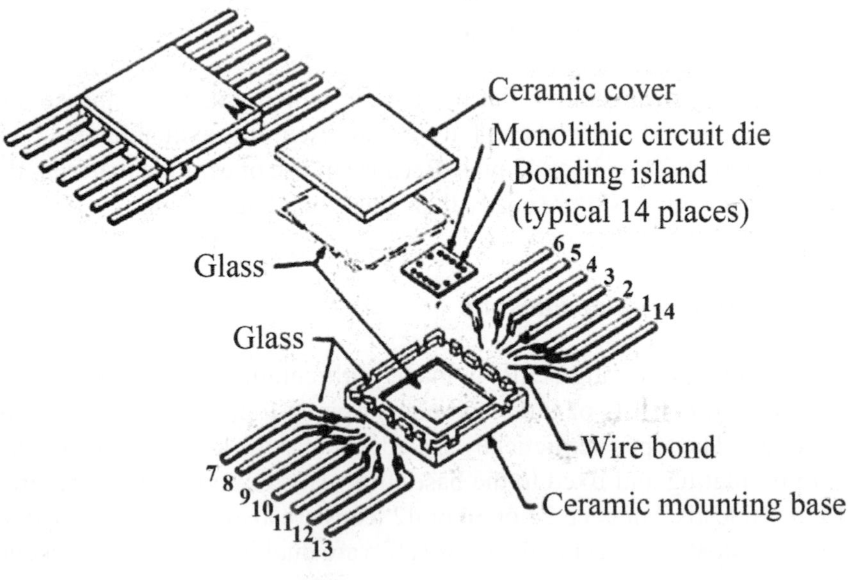

(b)

FIGURE 7.28 (a) Lead to 5 glass metal package and (b) 14 lead ceramic flat package.

KEYWORDS

- **oxidation**
- **high-k materials**
- **diffusion**
- **annealing**
- **deposition techniques**
- **sputtering**
- **lithography**
- **metallization**
- **evaporation**
- **etching**

REFERENCES

Chem, K. S.; Ayon, A.; Zhang, X.; Spearing, S. M. Effect of Process Parameters on the Surface Morphology and Mechanical of Silicon Structures After Deep Reactive Ion Etching (DRIE). *IEEE J. Micromech. Syst.* **2002,** *11* (3), 264–275.

Feridun, A. Siliconoxinitride Layers for Applications in Optical Waveguides. MS Thesis, Bilkent University, 2000.

George, J. *Preparation of Thin Films*; Macel Dekker Inc.: New York, 1992.

Kawachi, M.; Yasu, M.; Edahiro, J. Fabrication of SiO_2-TiO_2 Glass Planaroptical Waveguide by Flame Hydrolysis Deposition. *IEE Electron. Lett.* **1983,** *19*, 583–584.

Kim, Yun S.; Shin, D. W. Compositional Analysis of SiO_2 Optical Film Fabricated by Flame Hydrolysis Deposition. *J. Ceram. Process. Res.* **2002,** *3* (3), 186–191.

Lee, K.; Yu, D. K.; Chung, M.; Kang, J.; Kim, B. New Collector Undercut Technique Using a SiN Sidewall for Low Base Compact Resistance in InP/InGaAs SHBTs. *IEEE Trans. Electron Devices* **2002,** *44* (6), 1089–1082.

Madou, M. *Fundamentals of Microfabrication*; CRC Press: New York, 1997.

Marxer, C.; Christiaqn, T.; Gretillat, M. A.; Rooij, N. F. D.; Anthamatten, D.; Walk, B.; Vogel, P. Vertical Mirror Fabricated by Deep Reactive Ion Etching. *IEEE J. Micromech. Syst.* **1997**, *16* (3), 177–285.

Pandhumsoporn, T.; Feldbaum, M.; Gadgil, P.; Puech, M.; Maquim, P. High Etch Rate Anistropic Deep Silicon Plasma Etching for Fabrication of Microsensors. *In Proc. Micromach. Microfabr. Process Technol. II* **1996,** *2879*, 94–102.

Sahu, P. P. *VLSI Design*; McGraw Hill India: New Delhi, 2013.

Vasile, M. J.; Biddick, C.; Schwalm, S. Microfabrication by Ion Milling: the Lathe Technique. *J. Vac. Sci. Technol.* **1994,** *B12*, 2388–2393.

William, K. R.; Gupta, K.; Wasilik, M. Etch Rates for Micromachining Processing—Part II. *IEEE J. Micromech. Syst.* **2003,** *12* (6), 761–778.

CHAPTER 8

Tunnel FET: Working, Structure, and Modeling

SRIMANTA BAISHYA

Department of Electronics and Communication Engineering, National Institute of Technology, Silchar 788010, India
E-mail: s.baishya@yahoo.co.in

ABSTRACT

This chapter proposes a drain current model for tunnel FET using 2D Schrödinger's wave equation. The process for solving Schrödinger–Poisson equation with a potential function dependent on the position within the channel is described. The effect of temperature is also investigated. This proposed device has been recognized in digital inverter circuit and compared with conventional tunnel FET.

8.1 INTRODUCTION

With scaling, the subthreshold swing (SS) of conventional MOSFETs (metal–oxide–semiconductor field-effect-transistor) cannot be diminished below 60 mV/dec at room temperature. In this context, steep slope transistors such as tunnel-FET (field effect transistor) are identified as most attractive devices. The tunnel FET ON current is restricted by interband quantum mechanical tunneling. One of the foremost shortcomings of tunnel FET is its ON current limitation. To overcome these main difficulties, it is much suitable to use gate-all-around structure so that tunneling current can be increased up to ITRS requirement (ITRS, 2009). In this chapter, hetero-gate-all-around structure is used. In addition, gate-all-around tunnel FETs are most relevant for some of the very advantageous device implementations (Conzatti et al., 2012). In literature, the quantum wire wave functions were solved for constant potential function (Hanson, 2008). But, in reality, the potential profile depends on the position. Here, the method for solving Schrödinger–Poisson equation, with a potential function, which depends on the position within the channel, is described. Using 2D Schrödinger's wave equation, the models for the drain current are developed. Further, the effect of temperature is investigated. The proposed device can be realized in digital inverter and the digital performance parameters are compared with conventional tunnel FET.

8.2 WORKING PRINCIPLE

For a tunnel FET, the ON current is proportional to the electron/hole transmission probability $T(E)$ in the band-to-band tunneling (BTBT) mechanism, which is given by Knoch and Appenzeller (2005):

$$T(E) = \exp\left(-\frac{4\sqrt{2m^*}E_g^{\frac{3}{2}}}{3q\hbar\left(E_g + \Delta\phi\right)}\sqrt{\frac{\varepsilon_{si}t_{ox}t_{si}}{\varepsilon_{ox}}}\Delta\phi \right) \tag{8.1}$$

where m^* is the carrier effective mass, q is the electron charge, E_g is the bandgap, $\Delta\phi$ is the energy range over which tunneling can take place, and t_{ox}, t_{si}, ε_{ox} and ε_{si} are the oxide and silicon film thickness and dielectric constants, respectively. From eq 8.1 we have,

$$\lambda = \sqrt{\frac{\varepsilon_{si}t_{ox}t_{si}}{\varepsilon_{ox}}} \tag{8.2}$$

where the parameter λ has several different names, including screening length, natural length, or Debye length, and refers to the spatial extent of the electric field, or the length over which an electric charge has an influence before being screened out by the opposite charges around it (Streetman and Banerjee, 2000). It can be expressed in terms of the dielectric constants and thicknesses of the gate dielectric and semiconductor body of a device, and depends upon gate geometry. There are four important conditions in order for band-to-band tunneling to take place: available states to tunnel from, available states to tunnel to, an energy barrier that is sufficiently narrow for tunneling to take place, and conservation of momentum (Sze, 1981). This equation shows decreasing oxide thickness (t_{ox}), increasing oxide constant (ε_{ox}), and reducing bandgap (E_g), enhance the performance of the device. From the above equation, it can be seen that the band-to-band tunneling current, I_{B2B} increases exponential with an increase in V_{GS} as $T(E)$ increases. Boucart and Ionescu (Boucart and Ionescu, 2007) have proposed the use of high-k materials as thc gate dielectric high ε_{ox} in eq 8.1 in order to increase ON current (I_{on}). In this work, I_{on} enhancement has been done by modulating the bandgap (E_g).

8.3 DEVICE STRUCTURE

A 3D P-I-N gate all around cuboidal tunnel FET operating under reverse bias condition is shown in Figure 8.1. In this particular device, we have hetero-gate dielectric. Low-k gate oxide (SiO_2) whose dielectric constant is 3.9, is used at the drain side of the device and a high-k gate oxide (HfO_2) (Maity et al., 2014; 2014, 2016, 2017, 2018, 2019), having dielectric constant 25, has been applied at the source side of the device. The metal gate, with work function 4.5 eV, is used. A low band gap compound semiconductor $\text{InAs}\left(E_g = 0.36\text{eV}\right)$ is used as source to increase the tunneling probability.

The device has 20 nm channel length, 25 nm gate length with width = height = 5 nm. Also length of both the source and drain is 20 nm. Both oxide thicknesses are 1 nm. The length of both the oxides is 30 nm. The gate is applied over the entire channel along with a part over the source. The doping concentration of source, channel, and drain are 10^{21}, 10^{17}, and 5×10^{19} cm^{-3}, respectively. A high doping in the source side is used to enable easy tunneling from the source to the intrinsic region. The device has been operated by connecting drain to 0.7 V, source is grounded, and 1 V to the gate.

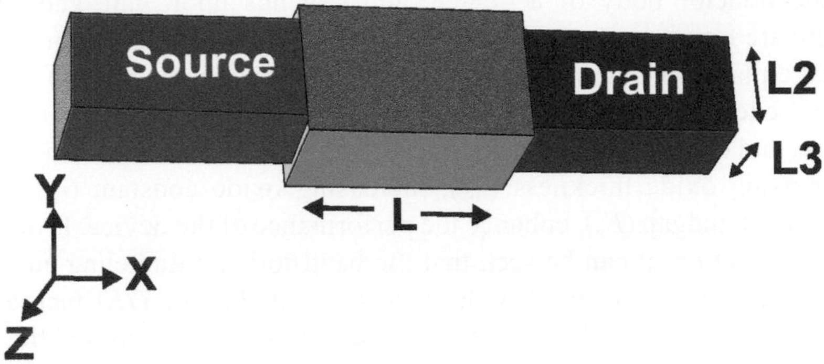

FIGURE 8.1 3D gate-all-around tunnel FET structure.

A modified 3D gate-all-around tunnel FET is shown in Figure 8.2, which is also simulated and results compared with the 3D cuboidal structure. On the source side, the high-k gate oxide HfO_2 and on the drain side the low-k gate oxide SiO_2 is applied. The gate covers the entire source and half of the channel length, that is, the device is a gate-all-around TFET where InAs channel and source is used. Band-to-band tunneling model is used. The feature of band gap narrowing is also activated. The doping concentrations in the source, channel, and drain regions are 10^{21}, 10^{16}, 10^{18} cm^{-3}, respectively.

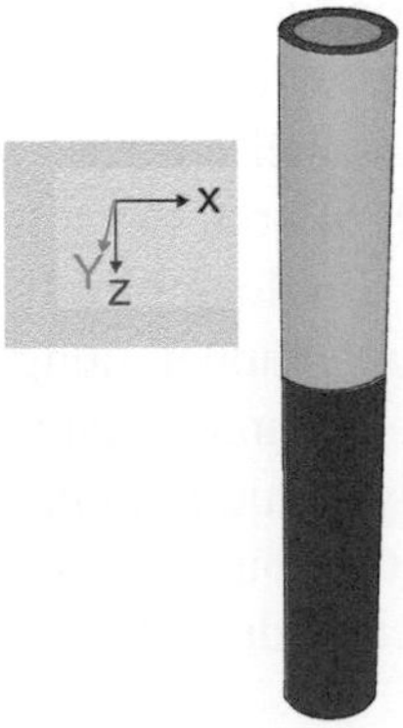

FIGURE 8.2 3D cylindrical tunnel FET structure.

8.4 SIMULATION

An effective mass of 0.223 m_o (InAs) is assumed for simulating the devices. Initially, InAs at source and channel is used, considering the proposed device in Figure 8.1. It is found that the drain current for InAs source and channel is better than only source with InAs. The simulated result shows a subthreshold swing of 42 mV/dec, whereas for InAs only in the source has 58 mV/dec as seen in Figure 8.3. This is due to the reason that InAs has lower band gap to provide better tunneling at source-channel junction when both source and channel are InAs.

Next, a device with optimized-for gate length is used and observed that such devices have better ON current at 40 nm channel length, but simultaneously the leakage (OFF) current is increased. The 30 nm channel is better in terms of OFF state leakage as well as SS, as shown in Figure 8.3. The SS is found to be 42 mV/dec and the OFF state leakage is of the order of $2.99\times10_{10}$ A/μm. Figure 8.4 shows the simulation of the device characteristics using Gaussian doping profile. In Gaussian doping profile, there is a non-negligible reduction in the ON-current for standard deviation greater than approximately 4. Clearly, it is observed that the Gaussian doping profile induces a smoother sub-band bending at the junction between the source and the channel and, hence, the ON current is decreased with increased standard deviation. Abrupt doping profile is compared with the Gaussian doping profile. The peak position of Gaussian doping in the channel is optimized. Finally, it is observed that when the peak position is far away from the source channel junction, we get an improved ON current but OFF current almost remains the same as seen in the Figure 8.5. Due to Gaussian doping in the channel, the band-to-band tunneling is reduced but the SS and OFF current can be improved.

Capacitance C_{gg}, C_{gd}, and C_{gs} are also obtained using the AC simulation. The capacitance C_{gd} is dominant during ON and OFF states as mentioned in Mookerjea et al. (2009). As observed from Figure 8.6, the capacitance C_{gd} of the proposed device is significantly less. Hence, miller capacitance is reduced. This is due to the fact that gate aligned over the channel and a part of the source.

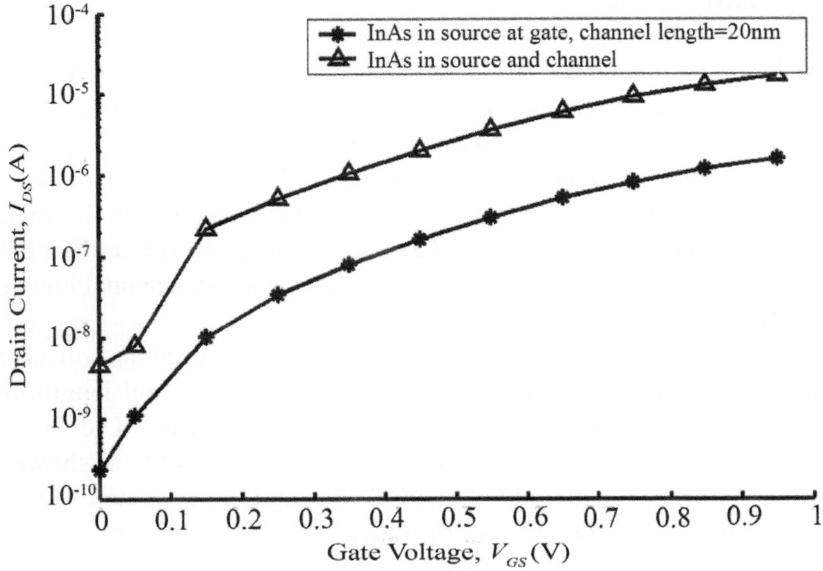

FIGURE 8.3 Comparison of InAs in source only, and in source and channel.

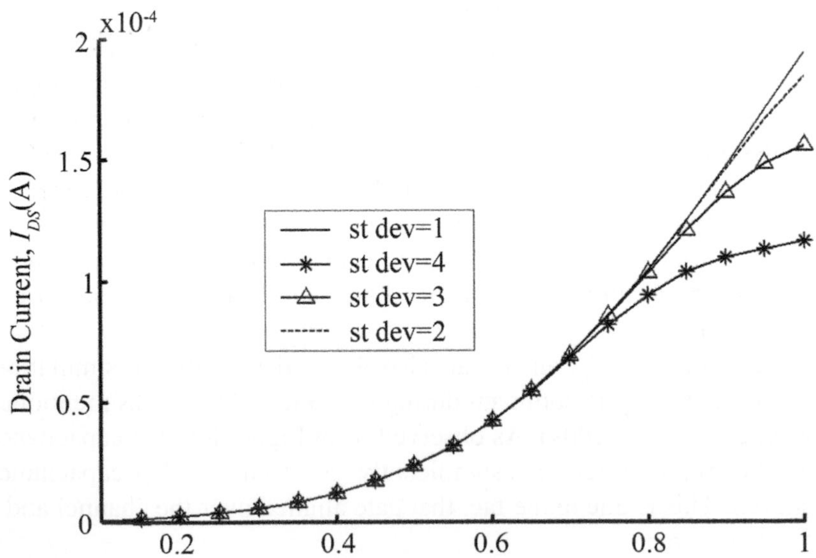

FIGURE 8.4 Comparison of I_D–V_G curves at various standard deviations.

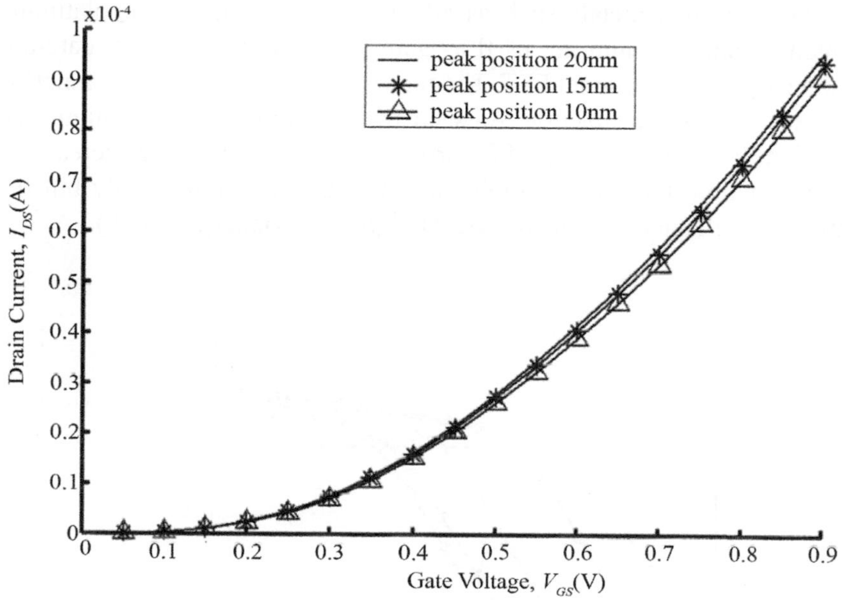

FIGURE 8.5 I_{DS}–V_{GS} characteristics for various peak positions.

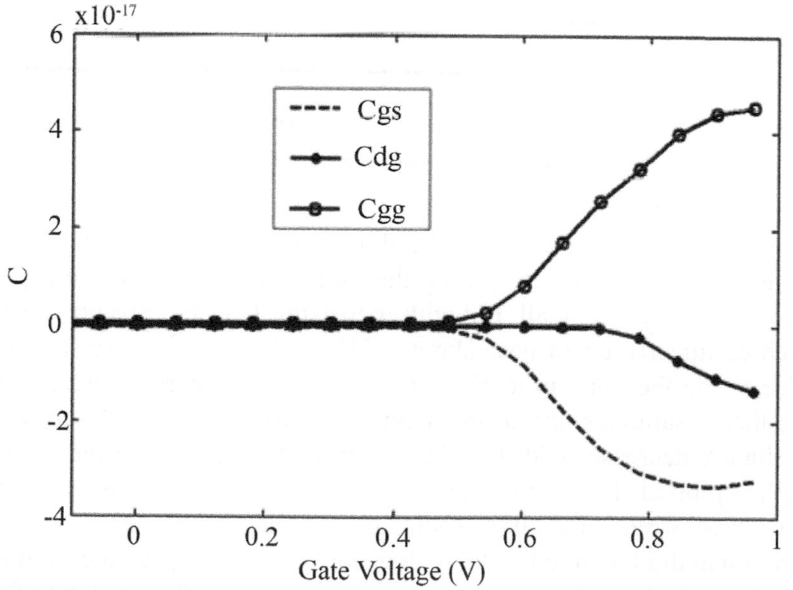

FIGURE 8.6 Capacitance voltage characteristics of 3D gate-all-around tunnel FET.

Also, various metals such as silver, gold, cobalt, nickel, platinum, tantalum, and titanium are used as gate materials for partially gate-all-around cylindrical Tunnel FET. The threshold voltage is a strong function of metal gate work function. As seen from Figure 8.7, for titanium and tantalum both the ON and OFF currents are better. With the increase of work function, the OFF and ON currents decrease significantly but the threshold voltage (V_T) is increased (Dadgow and Banerjee, 2008).

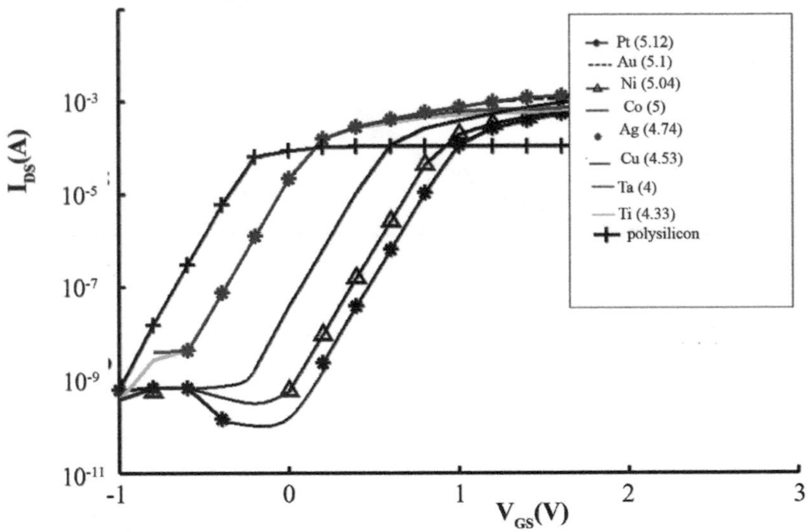

FIGURE 8.7 I_D–V_{GS} curves for various gate work functions.

In case of 3D Tunnel FET, the dominant component of C_{gg} is the gate-to-source capacitance. During the OFF state, V_{GS} = 0V, the capacitances are negligibly small. But with increasing V_{GS}, the capacitive effect increases linearly up to near about 0.5V as shown in Figure 8.8. This is due to the fact that up to V_{GS} = 0V, the channel charge increases and after that it saturates and to maintain the charge level almost constant, C_{gg} slightly decreases with V_{GS}. The potential function along the channel length is plotted. The entire gate capacitance is due to the gate-to-source capacitance, while the drain-to-source capacitance is very less. The TFET operation in digital circuits mainly depends on miller capacitance and this capacitance is dependent on C_{gd}. But in this GAA TFET, the effect of C_{gd} is negligible. Hence, better performance can be obtained.

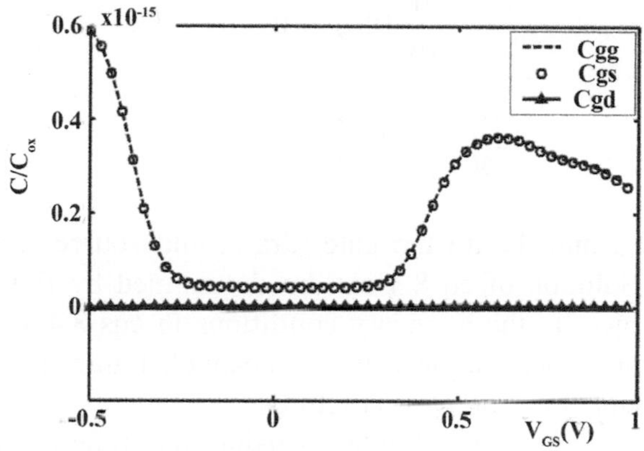

FIGURE 8.8 C–V characteristics for 3D tunnel FET.

8.5 MODEL DESCRIPTION

We have Poisson's equation (Bardon et al., 2010),

$$\frac{\partial^2 V}{\partial x^2}+\frac{\partial^2 V}{\partial y^2}=-\frac{\rho}{\varepsilon} \tag{8.3}$$

where $V(x,y)$ is the potential profile within the cuboid, ε is the permittivity of device material, and ρ is the charge density. The boundary conditions are given by:

$$V\left(x,L_1\right)=V_G-\frac{\phi_G}{q} \tag{8.4}$$

$$V\left(x,0\right)=V_G-\frac{\phi_G}{q} \tag{8.5}$$

$$V\left(x,0\right)=V_G-\frac{\phi_G}{q} \tag{8.6}$$

$$V\left(0,y\right)=-\frac{\phi_S}{q} \tag{8.7}$$

$$V(L,y)=V_{DS}-\frac{\phi_D}{q} \tag{8.8}$$

$$\frac{\partial V}{\partial x}(0,y)=0 \tag{8.9}$$

$$\varepsilon\left.\frac{\partial V}{\partial y}\right|_{y=0}-\varepsilon\left.\frac{\partial V}{\partial y}\right|_{y=L_2}=\frac{qn}{L_2} \tag{8.10}$$

where Φ_G, Φ_D, and Φ_S are the gate, drain, and source workfunctions, respectively. Solution of eq 8.3 can be determined by finite difference technique, subject to the boundary conditions in eqs 8.4–8.10 for some known electron concentration n in the channel. Finite differencing has flexibility in modeling complex structures.

The carrier density is calculated using the time-independent 2D Schrödinger's equation given by Merzbacher (1998)

$$\varepsilon\left.\frac{\partial V}{\partial y}\right|_{y=0}-\varepsilon\left.\frac{\partial V}{\partial y}\right|_{y=L_2}=\frac{qn}{L_2}$$

$$\frac{\partial^2\psi}{\partial x^2}+\frac{\partial^2\psi}{\partial y^2}=-\frac{2m}{\hbar^2}\{E-U(x,y)\}\psi(x,y) \tag{8.11}$$

where $\psi(x,y,E)$ is the wave function of the carriers with total energy E, effective mass m traveling under local effective potential $U(x,y)$, given by

$$U(x,y)=-qV(x,y)-\chi \tag{8.12}$$

where $V(x, y)$ is the potential function and χ is the electron affinity. Using the separation of variable technique, we can write the solution of eq 8.12 as:

$$\frac{1}{\psi_x(x)}\frac{\partial^2\psi_x(x)}{\partial x^2}=k_x{}^2 \tag{8.13}$$

$$\frac{1}{\psi_y(y)}\frac{\partial^2\psi_y(y)}{\partial y^2}=k_y{}^2 \tag{8.14}$$

$$\text{Now}, k_x{}^2+k_y{}^2+\frac{2m}{\hbar^2}[E-U(x,y)]=0 \tag{8.15}$$

$$k_x{}^2 = -k_y{}^2 - \frac{2m}{\hbar^2}[E - U(x,y)] = -k_p{}^2 \tag{8.16}$$

The effective mass m of the electron is assumed to be 0.98 m_0, where m_0 is the free electron mass. For a particular value of the local effective potential obtained from the solution of eq 8.3, the 2D wave equation can be formulated. The boundary conditions for wave function are

$$\begin{aligned} \psi(x,0) &= 0 \\ \psi(x,L_2) &= 0 \\ \psi(L,y) &= 0 \end{aligned} \tag{8.17}$$

In order to write the above boundary conditions, we assumed that the wave function inside the gate oxide and inside the substrate is zero. Moreover, at the drain there is no tunneling.

Using these boundary conditions, the solution of the wave function $\psi(x,y)$ can be written as

$$\begin{aligned} &\psi(x,y) = A\sin ky(C\sinh kx + D\cosh kx) \\ &k_x = k, k_p = jk \end{aligned} \tag{8.18}$$

Applying the boundary conditions, we get

$$\begin{aligned} &\psi(L,y) = 0 \text{ gives,} \\ &C\sinh kL + D\cosh kL = 0 \\ &X(x) = \frac{C}{\cosh kL}[\cosh kL \sinh kx - \sinh kL \cosh kx] \\ &= A_3 \sinh k(x-L) \end{aligned} \tag{8.19}$$

Also from $\psi(x, L2) = 0$, we get

$$k = \frac{n\pi}{L_2}, n = 1,2,3... \tag{8.20}$$

The wave function can be written as

$$\begin{aligned} &\psi_n(x,y) = C_n \sinh\frac{n\pi}{L_2}(x-L)\sin\frac{n\pi}{L_2}y,\ C_n = AA_3 \\ &\psi_n(0,y) = -C_n \sinh\frac{n\pi}{L_2}L\sin\frac{n\pi}{L_2}y = \psi_o \end{aligned} \tag{8.21}$$

The constant C_n is given by

$$C_n = \begin{cases} -\dfrac{4\psi_o}{n\pi \sinh(n\pi L/L_2)}, & n \text{ odd} \\ 0, & n \text{ even} \end{cases} \tag{8.22}$$

The final solution of wave function is given by

$$\psi(x,y) = \frac{4\psi_o}{\pi} \sum_{n=\text{odd}}^{\infty} \frac{\sinh[n\pi(L-x)/L_2]}{n \sinh(n\pi L/L_2)} \sin\left(\frac{n\pi}{L_2} y\right) \tag{8.23}$$

The 2D electron concentration can be determined with the help of numerical integration, and is given by

$$n(x,y) = \int_0^{Ec} |\psi(x,y)|^2 \, dE \tag{8.24}$$

The probabilistic current density is given by (Mookerjea et al., 2009)

$$J(x,y,z) = \frac{\hbar}{m} \mathrm{Im}(\psi^* \nabla \psi) \tag{8.25}$$

8.6 MODEL VERIFICATIONS

2D Schrödinger's model is validated for 20, 30, and 40 nm channel devices. The optimized gate length of the device for improved ON current is found to be 40 nm. But, it has increased OFF state leakage current. The model is tested for transfer and drain characteristics of the proposed device as well. Furthermore, it is also tested for gate-length scaling. The model is tested on a gate-all-around device with 1 nm oxide thickness. OFF current and subthreshold slope are the two important parameters. At the OFF state, for 40 nm channel length, OFF current increases and also the subthreshold slope is less.

8.7 TEMPERATURE EFFECT ON ELECTRICAL PARAMETERS

Here, effect of temperature on tunnel bandgap and SS is discussed.

8.7.1 Effect of Temperature on the Tunnel Bandgap

The reduction of the energy bandgap or the tunneling bandgap at elevated temperature can be understood from:

$$E_g(T) = E_g(0) - \frac{\alpha T^2}{T + \beta} \tag{8.26}$$

where α and β are material constants and $E_g(0)$ is the limiting value of E_g at 0° K. Hence, with increasing temperature the tunneling bandgap should reduce. Figure 8.9 shows the variation of the tunneling bandgap with temperature for the conventional TFET as well as the proposed device. The Si–Ge has a smaller bandgap than silicon because of larger lattice constant and strain.

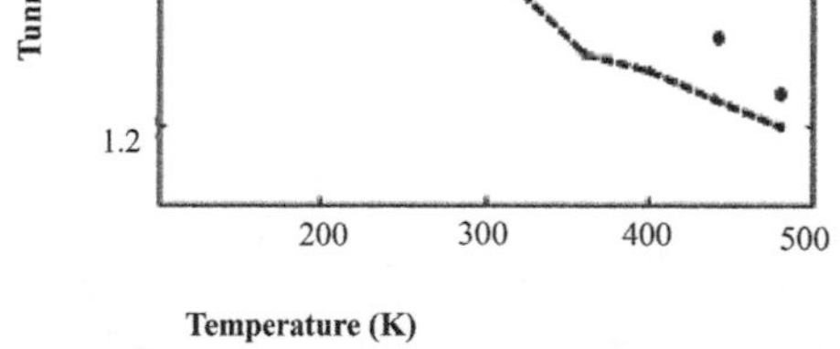

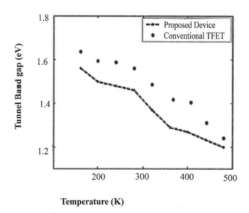

FIGURE 8.9 Tunneling bandgap with respect to temperature.

As expected, it is evident from Figure 8.9 that for the proposed device the variation in tunneling bandgap with temperature is much lesser compared to that of the conventional devices. It is observed that the reduction in tunneling gap with temperature is more for the conventional TFET, and hence, the proposed device is less dependent on temperature. Since raised buried oxide in drain with Si–Ge layer in the source side provides less electric field, we get a higher carrier mobility with lesser channel resistance.

8.7.2 Effect of Temperature on Subthreshold Swing

SS is defined as the amount of V_{GS} needed to change the I_{DS} by one decade, and is given by

$$SS = 2.3\left(\frac{E + A'}{E^2}\frac{dE}{dV_{GS}}\right)^{-1} \tag{8.27}$$

where $A' = DEg^{3/2}$, E is the magnitude of the electric field, V_{GS} is the gate-to-source voltage, D is the parameter that depends on the effective mass in the valence and conduction bands, and E_g is the gap energy.

From eq 8.27, it is observed that SS is a function of A', which itself is a function of energy bandgap E_g. With increasing temperature, the tunneling bandgap reduces, hence A' decreases too. As a result, the SS is increased.

In Figure 8.10, it is observed that the SS values are decreasing with increasing tunnel bandgap. Comparing conventional device and proposed device, it is seen that change in SS value, with temperature, is more in conventional device. Moreover, the proposed device has a SS which is much lower than the conventional device at room temperature.

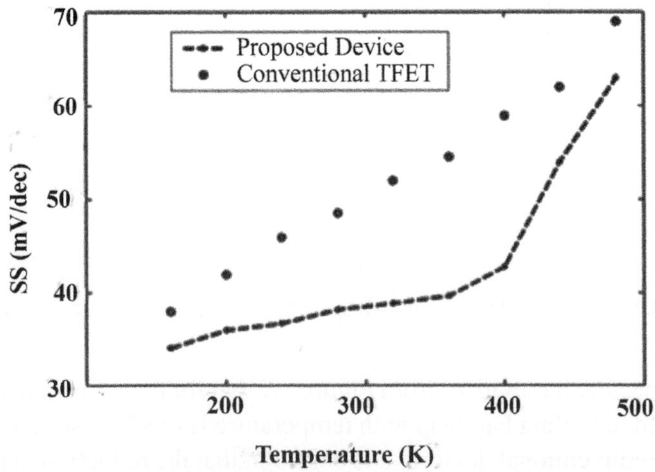

FIGURE 8.10 Subthreshold swing enhances with the decrease in tunnel bandgap.

8.8 DIGITAL CIRCUIT APPLICATIONS

Recently, leakage reduction using steep subthreshold transistor gained great attention. A step subthreshold transistor (TFET) can be operated at very low threshold voltages with ultra low leakage and low supply voltages. TFETs are found to be extremely power efficient in logic circuit applications. The dynamic power dissipation for an inverter is given by

$$P_D = fCV_{DD}^{\ 2} \tag{8.28}$$

where f is the frequency at which the inverter is switched on, C is the total gate capacitance, and V_{DD} is the power supply. The frequency is related with propagation delay, lower the propagation delay higher is the frequency at which the circuit can be operated, of course, with higher power dissipations as well. A figure of merit or a quality measure of the particular circuit technology is power delay product (PDP), given by

$$\text{PDP} = P_D(t_{PHL} + t_{PLH}) \tag{8.29}$$

where t_{PHL} and t_{PLH} are turn OFF and turn ON delay time. In Table 8.1, the electrical parameters of conventional TFET and the gate-all-around TFET are compared.

TABLE 8.1 Comparison of Gate-All-Around Tunnel FET Structure with Conventional Tunnel FET.

	SS (mV/Dec)	OFF current (A)	ON current (A)	V_T (V)	Total capacitance (Cgg)	Transconductance g_m(S)	Leakage power (µW/µm)	Dynamic power (µW/µm)	Intrinsic device delay (sec)	Supply voltage (V)
Si TFET	68	10^{-14}	1×10^{-5}	0.7	0.69	6.25×10^{-6}	0.5×10^{-8}	3.483	62P	0.5
Gate-all-around TFET	48	7×10^{-9}	0.99×10^{-3}	0.12	0.54	1.899×10^{-3}	3.5×10^{-3}	2.384	3P	0.5

KEYWORDS

- **TFET**
- **subthreshold swing**
- **gate-all-around TFET**
- **C-V characteristics**
- **temperature effect**

REFERENCES

Bardon, M. G.; Neves, H. P.; Puers, R.; Van Hoof, C. Pseudo Two Dimensional Model for Double-Gate Tunnel FETS considering Junction Depletion Regions. *IEEE Trans. Electron Devices* **2010,** *57* (4), 827–834.

Boucart, K.; Ionescu, A. Double gate Tunnel FET with High-k Gate Dielectric. *IEEE Trans. Electron Devices* **2007,** *54* (7), 1725–1733.

Conzatti, F.; Pala, M. G.; Esseni, D.; Bano, D.; Selmi, L. Stain-Induced Performance Improvements in InAs Nanowire Tunnel FETs. *IEEE Trans. Electron Devices* **2012,** *59*, 2085–2092.

Dadgow, H.; De, V.; Banerjee, K. In *Statistical Modeling of Metal Gate Workfunction Variability in Emerging Device Technology and Implication for Circuit Design*, Proceedings of the IEEE International Conference on Computer Aided Design, 2008.

Hanson, G. W. *Fundamentals of NANO Electronics*; Pearson Education: USA, 2008.

Knoch, J.; Appenzeller, J. A Novel Concept for Field-Effect Transistors—The Tunneling Carbon Nanotube FET. *IEEE Device Research Conf. Digest* **2005,** 153–156.

Maity, N. P.; Maity, R.; Thapa, R. K.; Baishya, S. Study of Interface Charge Densities for ZrO_2 and HfO_2 Based Metal Oxide Semiconductor Devices. *Adv. Mat. Sci. Eng.* **2014,** *2014*, 1–6.

Maity, N. P.; Maity, R.; Thapa, R. K.; Baishya, S. A Tunneling Current Density Model for Ultra Thin HfO_2 High-k Dielectric Material Based MOS Devices. *Superlat. Microstruc.* **2016,** *95*, 24–32.

Maity, N. P.; Maity, R.; Baishya, S. Voltage and Oxide Thickness Dependent Tunneling Current Density and Tunnel Resistivity Model: Application to High-k Material HfO_2 Based MOS Devices. *Superlat. Microstruc.* **2017,** *111*, 628–641.

Maity, N. P.; Maity, R.; Baishya, S. A Tunneling Current Model with a Realistic Barrier for Ultra Thin High-k Dielectric ZrO_2 Material based MOS Devices. *Silicon* **2018,** *10* (4), 1645–1652.

Maity, N. P.; Maity, R.; Maity, S.; Baishya, S. Comparative Analysis of the Quantum FinFET and Trigate FinFET Based on Modeling and Simulation. *J. Comput. Electron.* [Online] 2019, doi.org/10.1007/s10825-018-01294-z.

Merzbacher, E. *Quantum Mechanics*, 3rd ed.; John Wiley & Sons: New York, 1998.

Mookerjea, S.; Krishnan, R.; Datta, S.; Narayanan, V. On Enhanced Miller Capacitance Effect in Interband Tunnel Transistors. *IEEE Electron. Device Lett.* **2009,** *30* (10), 1102–1104.

Semiconductor Industry Association (SIA), International Technology Roadmap for Semiconductors (ITRS) Report, 2009.

Streetman, B.; Banerjee, S. *Solid State Electronic Devices*, 5th ed.; Prentice Hall, Inc.: New Jersey, 2000.

Sze, S. M. *Physics of Semiconductor Devices*, 2nd ed; John Wiley & Sons, Inc.: New York, 1981.

CHAPTER 9

Heusler Compound: A Novel Material for Optoelectronic, Thermoelectric, and Spintronic Applications

D. P. RAI

Physical Sciences Research Center (PSRC), Department of Physics, Pachhunga University College, Aizawl 796001, India
E-mail: dibya@pucollege.edu.in

ABSTRACT

Heusler compounds are center of scientific research due to their exceptional magnetic and transport properties for the development of spintronic technology. Heusler compounds are the class of intermetallic materials with 1:1:1 (called Half-Heusler) and 2:1:1 (called full-Heusler) alloys comprising of more than 1000 combinations. The combination of half and full Heusler are given by XYZ and X_2YZ respectively. Here, in XYZ/ X_2YZ systems, X and Y are transition metals and Z is in the p-block. The higher negativity at Y site than at X site makes the full-Heusler compounds inverse or indirect one, otherwise they are called direct full-Heusler alloys. Most of the Heusler alloys crystallize in FCC structure with space group F-43m (#216) and Fm3m (#225). Half-Heusler and inverse Heusler alloys takes the space group of F-43m. While direct full-Heulsr alloys takes Fm3m space group. One of the most studied properties is coexistence of metal–semiconductor properties with ferromagnetic behavior. The presence of band gap in one of the spin channels at the Fermi level gives 100% spin polarization. The material which exhibit such properties are called half-metal ferromagnetic materials. They are promising for next generation spintronics technology. Other than spintronic, Heusler alloys have diverse fascinating properties, such as magnetoresistance, the Hall effect, low

Gilbert damping, high Curie temperature, ferro-, antiferro-, and ferrimagnetism, half- and semimetallicity, semiconductivity with spin filter ability, superconductivity, topological band structure etc. The magnetic behavior is studied in terms of a double-exchange mechanism between neighboring magnetic ions. Also the origin of semiconducting band gap is due to the d-d hybridization of transition elements which results in bonding and antibonding states. The half-metallicity and magnetic behavior of many Heusler compounds can be predicted by using Slater-Pauling rule which relies on the number of total valence electron count. The Slater–Pauling rule is given by, $M_t=Z_t-24/18$, where M_t is the total magnetic moment and Z_t is the total valence electrons. The finite integer value of M_t gives the ferromagnetic half-metal and 0 gives semiconducting behavior. The fractional value of M_t gives metallic behavior without any spin band gap. Other than half-metallicity a great interest has been attracted in the fields of thermoelectrics as well. Heusler alloys with M_t value equal to zero are semiconductors with narrow band gap and high value of melting temperature. This shows that Heusler alloys can operate up to high temperature range as a thermoelectric generator (TEG). Therefore, to understand the intriguing properties, the structural, the electronic structure, magnetic, and the thermoelectric properties have been investigated using density functional theory (DFT) as implemented in the Wien2K and Quantum Espresso package.

9.1 INTRODUCTION

A Heusler compound is a name given to an intermetallic compound after a German mineralogist and a physicist Friedrich Heusler. Heusler reported that few alloys comprise copper, manganese, and ferromagnetic element despite of paramagnetic constituent (Heusler, 1903). Its magnetism varies considerably with heat treatment and composition (Knowlton, et al. 1912). In 1934, Bradley and Rogers discovered the ferromagnetic phase at room temperature with fully ordered structure of $L2_1$ type (Bradley, 1934). This system has a primitive cubic lattice of copper atoms with alternate body centered by manganese and aluminum. In order to solve this ferromagnetism, further insight of the solid properties were closely analyzed from experiment as well as theory. For instance, the prototype Heusler compound Cu_2MnAl with fcc-$L2_1$ structure demonstrates a ferromagnetic configuration.

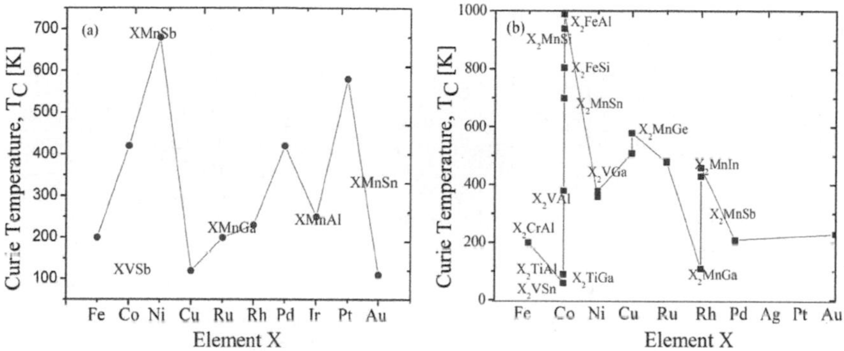

FIGURE 9.1 Curie temperature distribution of both (a) half- and (b) full-Heusler bulk alloys with respect to the element X (Adapted from Hirohata et al., 2006).

Heusler compounds have got substantial research attention due to their multifunctional properties for Spintronic (Wolf et al., 2010), optoelectronic (Umetsu et al., 2008; Strand et al., 2005; Dorpe et al., 2005), superconductivity (Waki et al., 1985; Ishikawa et al., 1982), shape memory (Ullakko et al., 1996; Kaufmann et al., 2016), thermoelectric (Tritt, 2011; Snyder et al., 2008; Rai et al., 2016; 2017), and topological insulator (Casper et al., 2012; Chadov et al., 2010; Lin et al., 2010; Xiao et al., 2005; Wang et al., 2016). Another pivotal property of Heusler compound for making innovative spin-device is their high Curie temperature (above room temperature) (Zutic et al., 2005; Wurmehlet al., 2005; Hirohata et al., 2006). The Curie temperature (T_C) of different Heusler compounds is shown in Figure 9.1.

9.2 TYPES OF HEUSLER COMPOUNDS

Heusler compounds were constructed taking the stoichiometric ratio as 2:1:1 or 1:1:1 by selecting the X, Y, and Z elements from the periodic table as highlighted in Figure 9.2. The former stoichiometric composition gives the full-Heusler, while the latter is called a half-Heusler compound. The chemical formulae for a full-Heusler and a half-Heusler are denoted as X_2YZ and XYZ, respectively.

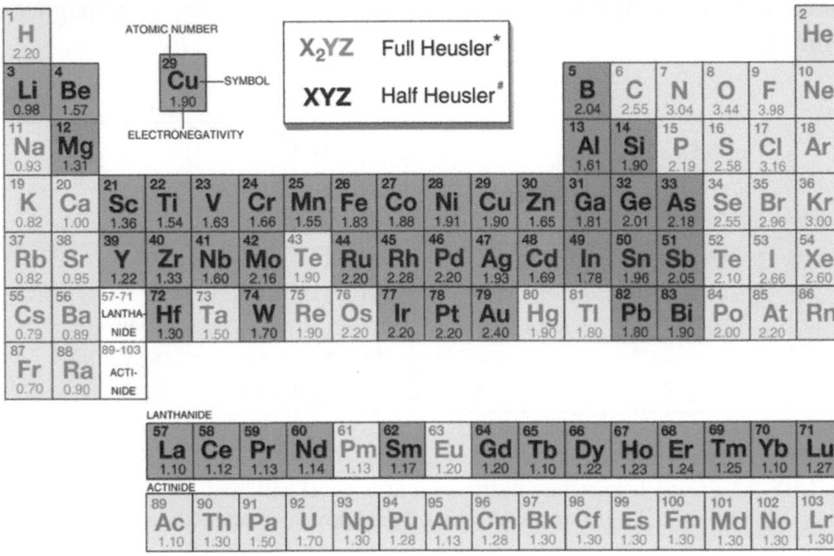

FIGURE 9.2 Periodic table of Heusler compounds.

FIGURE 9.3 The fully ordered $L2_1$ crystal structure and primitive cell of full-Heusler alloy (X = Blue, Y = Green, and Z = Red).

9.2.1 Full-Heusler Alloy (FHA)

A full-Heusler is an intermetallic quaternary compound comprises 4 FCC sub-lattices. A full-Heusler compound crystallizes in $L2_1$ structure with space group F-m3m (225). Chemical formula of full-Heusler compound is written as X_2YZ/XX'YZ with stoichiometric structure 2:1:1, where X/X' and Y are d-elements (Ni, Co, Fe, Mn, Cr, Ti, V, and so forth), and Z is III, IV, or V group elements (Al, Ga, Ge, AS, Sn, In, and so on). Here, X, Y, and Z atoms take the Wyckoff positions 4c (1/4, 1/4, 1/4), 4d (3/4, 3/4, 3/4), 4a (0, 0, 0), and 4b (1/2, 1/2, 1/2) individually. Figure 9.3 demonstrates a $L2_1$ structure of X2YZ along with primitive cell.

9.2.2 *Half-Heusler Alloy (HHA)*

Half-Heusler compounds are also called semi-Heusler compounds having stoichiometric ratio 1:1:1. Chemical formula is XYZ, with two prominent transition metals (TMs) X and Y. Whereas Z is a sp valence element, being the third constituents and crystallizes in cubic $C1_b$ structure (F43m, space-group-216) with one of the positions void. The semi-Heusler compounds would basically commensurate to full-Heusler compounds X_2YZ, as both structures were involved four inter-penetrating FCC lattices. The distinguishable feature in full-Heusler compounds are two sub-lattices occupied by the X atoms and void space. While on XYZ compounds, the three sub-lattices are filled up and one remains empty. Fundamentally, those semi-Heusler compounds frame a ternary-stuffed variation for conventional electron closed-shell semiconductors that take the zinc blende (ZnS-type) structure, for example GaAs. Those eight valence electrons would be distributed around three atoms rather than two. The third atom forms the octahedral symmetry in the ZnS-type cage. This normally prompts a casing of a rock salt-like sub-lattice depicting an ionic bond (Casper et al., 2012). The Wyckoff positions about XYZ half-Heusler compounds are 4d (3/4, 3/4, 3/4), 4a (0, 0, 0), and 4b (1/2, 1/2, 1/2) for X, Y, and Z atoms, respectively. When the last two atoms, Y and Z, are put at 4a (0, 0, 0) and 4b (1/2, 1/2, 1/2), at that point it gives rocksalt structure. The X atom is situated at the octahedral coordinate, situated at 4c (1/4, 1/4, 1/4) inside the cube leaving the other position 4d (3/4, 3/4, 3/4) unfilled. The Z ion takes the positions (0, 0, 0) making this structure equivalent to the zinc blende structure, a most commonly adopted by large semiconductor groups (Nanda et al., 2007). Despite the fact that it took around two decades to understand that it is realizable to make one of the four sub-lattices empty in X2YZ, yielding XYZ compound. The primary XYZ kind of half-Heusler compound was found in a single-phase alloy; CuMnSb is an example that takes structure similar to that of the full-Heusler (Castelliz, 1952). Nowotny et al. (1952) performed X-beam diffraction to confirm the atomic positions in semi-Heusler compounds, which takes the MgAgAs-type structure similar to the full-Heusler structure keeping one of the four sub-lattices empty. The connection between these distinctive crystal lattices is shown in Figure 9.4.

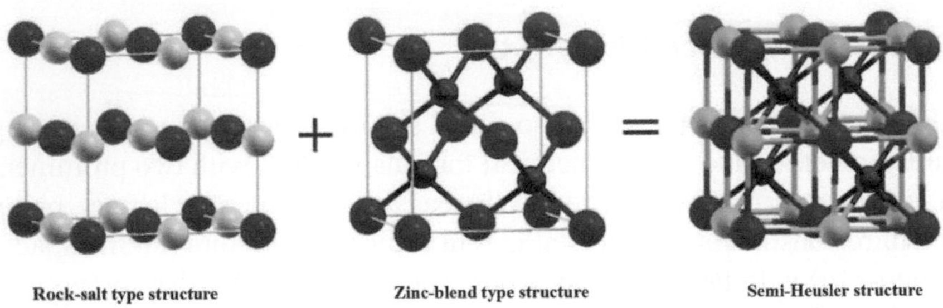

FIGURE 9.4 A schematic representation of rock salt, zinc-blend, semi-Heusler, and full-Heusler type structure.

9.3 THEORY

9.3.1 Theory and Calculation Details

The macroscopic observable physical phenomena of a solid are described by understanding the distribution of atomic positions and electron clouds in an atomic scale. Density functional theory (DFT) is the most reliable theory to study periodic solids that provided us the physics underlying different phenomena depending on the single potential model. In crystalline solids, there are a number of interacting atoms under a crystal potential (effective potential). However, we describe systems on the basis of a number of nuclei and interaction of electrons through the Coulombic (electrostatic) forces. In order to understand the structural trend and magnetism of a solid, in relation to macroscopic properties, it is necessary to scan the material in its atomistic scale. It is one of the most widely used techniques in computational condensed matter physics, developed by Kohn, Hohenberg, and Sham (Hohenberg et al., 1964; Kohn et al., 1965). This theory provides a modern tool to study the ground state properties of atoms, molecules, and solids. The principle of DFT enables to reduce an interacting many-body system of fermions into a single body problem by using electron density instead of many-body wave functions. This theory relies on the Kohn–Sham equations of this assisting non-interacting system.

$$\left[-\frac{\hbar^2}{2m}\nabla^2 + V_S(\vec{r})\right]\phi_i(\vec{r}) = \in_i \phi_i(\vec{r}) \tag{9.1}$$

which produces the ϕ_i orbital that reproduces the density $n(\vec{r})$ of the authentic many-body system,

$$n(\vec{r}) = n_S(\vec{r}) = \sum_i^N |\phi_i(r')|^2 \tag{9.2}$$

The effective single-particle potential can be written in more detail as:

$$V_S(\vec{r}) = V(\vec{r}) + \int \frac{e^2 n_S(\vec{r}')}{|\vec{r} - \vec{r}'|} d^3r' + V_{XC}[n_S(\vec{r})] \tag{9.3}$$

The second term in the eq 9.3 stands for the so-called Hartree expression with the electron–electron Coulomb repulsion, while the last expression V_{XC} is called the exchange correlation potential. Now, the Hartree expression and V_{XC} depend on $n(\vec{r})$ that depends on the ϕ_i which in turn relies on $\hat{V}_S$. Now, the problem is that the Kohn–Sham equation has to be solved in a self-consistent way. One typically begins with a first guess $n(\vec{r})$, then works out the corresponding $\hat{V}_S$ and solves the Kohn–Sham equations for the ϕ_i. The techniques in DFT are complex and different, and can be understood by considering the following approaches: (1) the last term of eq 9.3 has to be approximated with the electron density including the effect of gradient correction factor. (2) A gradient that measures the rate of change of some property. In this case, the gradient arises with the slow variation of the electron density cloud, and is known as generalized gradient approximation (GGA) as described by Perdew, Burke, and Ernzerh of (PBE) (Perdew et al., 1956). The effective potential now can be written as:

$$V_{eff}(r) = V_{ext}(r) + e^2 \int \frac{n(r^{'})}{\left|r - r^{'}\right|} dr^{'} + \frac{\delta E_{XC}[n]}{\delta_n} \tag{9.4}$$

The complete mathematics of this theory with varying exchange potential is programmed in many computational packages like SIESTA (José et al., 2002), WIEN2k (Blaha et al., 2001), Quantum Espresso (Giannozzi et al., 2009), Abinit (Gonze et al., 2002), ELK (Dewhurst et al., 2011), Exciting (Gulans et al., 2014), Castep (Kristallogr, 2005), VASP (Kresse, 2010) etc. In our case, we have opted two packages (WIEN2k and Quantum Espresso) for calculation. All the related physical and chemical properties of Heusler compounds were calculated by the full potential linearized augmented plane wave (FLAPW) and ultrasoft potential method.

9.3.2 Structural Properties

The minimum energy (*E*) of a system is calculated by varying its lattice constant. A unit cell volume corresponds to this minimum energy gives the most stable structure. A theoretical lattice constant and bulk modulus can be determined from the relaxed unit cell volume. The smooth curve of energy versus volume can be obtained by fitting the calculated total energy to Murnaghan's equation of state (Murnaghan, 1944). A series of total energy calculations as a function of volume can be fitted to an equation of states according to Murnaghan.

$$E(V) = E_0 + \left[\frac{(V_0/V)^{B_0^{'}}}{B_0^{'} - 1} + 1 \right] - \frac{B_0 V_0}{B_0^{'} - 1} \tag{9.5}$$

where E_0 is the equilibrium energy at $T = 0$ K, B_0 is the bulk modulus, and $B_0^{'}$ is pressure derivative of the bulk modulus at the equilibrium volume. Pressure is $P = -\frac{dE}{dV}$, and Bulk modulus is $B_0 = -V\frac{dP}{dV} = V\frac{d^2E}{dV^2}$. As the experimentally determined alloys are not optimized at their ground state, therefore, the optimized lattices were obtained at 0 K. The structural optimization was tested for the ferromagnetic (FM) and non-magnetic (NM) configuration. The structure that corresponds to a minimum energy gives the ground state stable structure. In Figure 9.5, structural optimization is performed by minimizing the total energy with respect to unit cell volume. The equilibrium values of lattice parameters a (Å) for some of the half-Heusler and full-Heusler compounds are summarized in Table 9.1. In all calculations, the calculated lattice parameters are in qualitatively close agreement with the experimental data. The remaining calculations have been performed at these optimized values.

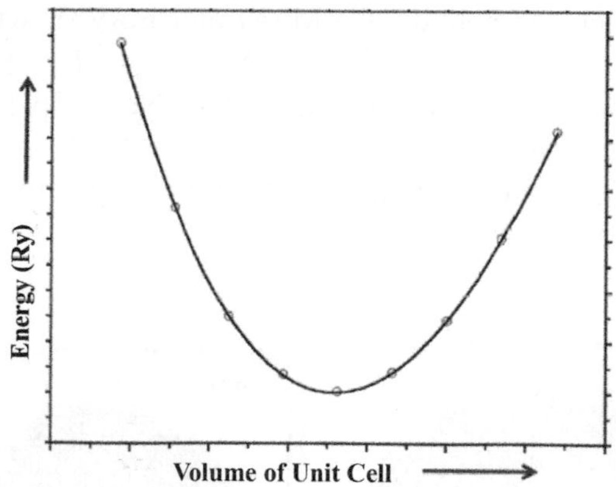

FIGURE 9.5 Total energy of arbitrary compound as a function of volume.

9.4 CHARACTERISTIC OF HEUSLER COMPOUND

Heusler alloys exhibit diverse characteristics depending on their compositions. The most common features of Heusler alloys arc (1) half-metal, (2) semiconductor, and (3) semi-metal. Figure 9.6 demonstrates the density of states (DOS) of half-metal, a normal metal, semiconductor, and semi-metal. An atomistic behavior of Heusler alloys can be analyzed by Slater Pauling rule. It is one of the necessary conditions of a Heusler alloy to examine the characteristics like metallic, semiconductor, or half-metal. It simply implies that the presence of electrons near Fermi level requires the numbers of charge carriers per unit cell ($Z\uparrow$ and $Z\downarrow$), where $Z\uparrow$ signifies the number of spin-up electrons and $Z\downarrow$ refers to number spin-down electrons. Both $Z\uparrow$ and $Z\downarrow$ are integer numbers and their difference is also an integer, resulting in an integer moment (Fecher et al., 2005).

The linear behavior of the total magnetic moment with the total valence electrons is an extension of the Slater–Pauling rule (Slater, 1936; Pauling, 1938; Galanakis et al., 2006). It states that the total number of electrons is

$Z_t = Z\uparrow + Z\downarrow$ in ferromagnets, where the Z_t is the number of spin-up ($Z\uparrow$) and spin-down ($Z\downarrow$) electrons per atom. The difference of $Z\uparrow$ and $Z\downarrow$ gives the total magnetic moment $M_t = Z\uparrow - Z\downarrow$. Replacing $Z\uparrow = M_t + Z\downarrow$ in ($M_t = Z\uparrow - Z\downarrow$) results in $M_t = (Z_t - 2Z\downarrow)\ u_B$/atom. Magnetic behavior of Heusler compounds (half or full) can be determined (Galanakis et al., 2006). $M_t = (Z_t - 24)\ u_B$/atom for the full-Heusler and $M_t = (Z_t - 18)\ u_B$/atom for the half-Heusler alloys (Galanakis et al., 2006). Let us consider ferromagnetic full-Heusler compounds, Co_2MnAl and Fe_2VAl, whose electronic configurations are {Co:[Ar]$4s^23d^7$}$_2$ {Mn:[Ar]$4s^3d^5$} {Al:[Ne]$3s^23p^1$} and {Fe:[Ar]$4s^23d^6$}$_2$ {V:[Ar]$4s^33d^3$} {Al:[Ne]$3s^23p^1$}, respectively. Co_2MnAl will have total electrons count $Z_t = 9 \times 2 + 7 + 3 = 28$ around the Fermi energy (E_F). Thus following the Slater Pauling's rule of full-Heusler compound, we get $M_t = (Z_t - 24) = 4.00\ u_B$/atom.

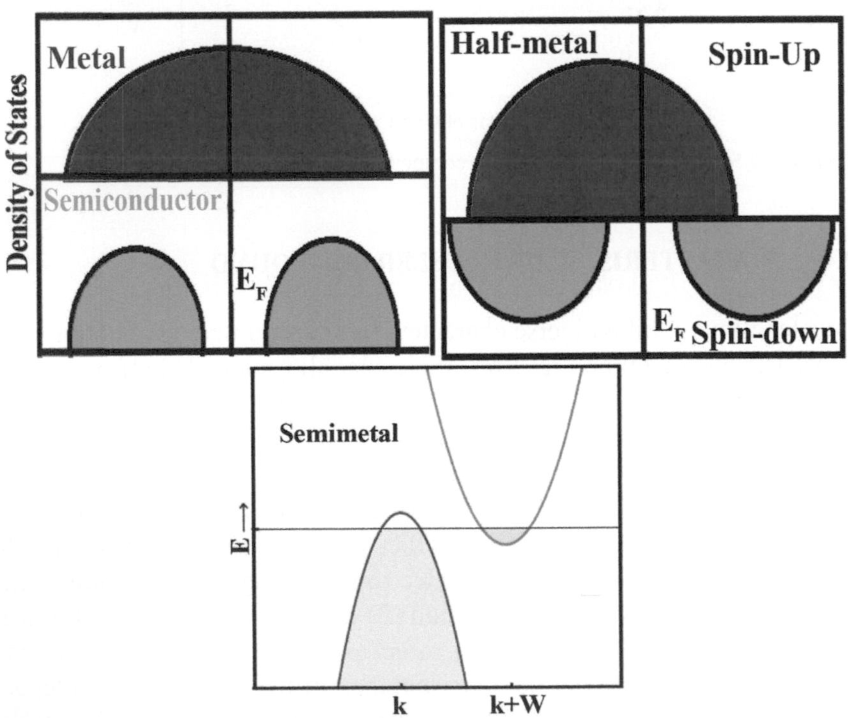

FIGURE 9.6 Schematic representation of the density of states for a half-metal compared to a normal metal, semiconductor, and semi-metal.

The finite integer value of magnetic moment is an evidence for these types of materials to exhibit half-metal ferromagnetic behavior. While in Fe_2VAl, $Z_t = 8 \times 2 + 5 + 3 = 24$, $M_t = (Z_t - 24) = 0.0\ u_B$/atom. The presence of magnetic moment equivalent to 0.0 u_B/atom gives NM behavior. Similar explanation also follows for half-Heusler compound with Slater–Pauling rule, $M_t = (Z_t - 18)\ u_B$/atom. Following this rule, we have sorted out a few magnetic and non-magnetic Heusler compounds as tabulated in Table 9.1. A selective number of Heusler compounds following the Slater–Pauling rule are also presented in Figure 9.7.

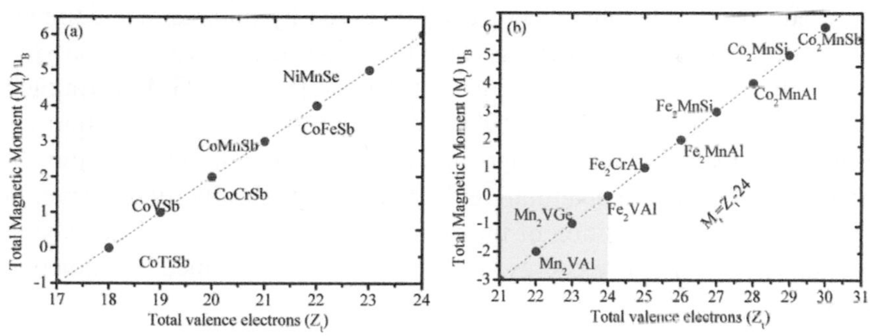

FIGURE 9.7 The expected total spin moments for the selected half-Heusler and full-Heusler alloys from Slater–Pauling rule. The dashed line represents the Slater–Pauling behavior (Adapted from Webster et al., 2006).

9.4.1 *Half-Metal (HM)*

Half-metallic behavior is reported in several materials like oxide perovskites (Ramirez, 2017) and double perovskites (Serrate et al., 2007), Fe_3O_4 (Pénicaud et al., 1992), and CrO_2 (Schwarz, 1986). However, from experimental point of view the concrete proof of nearly complete spin polarization has only found in $La_{2/3}Sr_{1/3}MnO_3$ at low temperatures (Park et al., 1998; Bowen et al., 2003). Unfortunately, all these materials show low Curie temperature below the room temperature that limits their application in technological devices. An attention has been shifted toward materials with high T_c. Heusler compound was discovered in 1903, but this material was not highlighted as an energy material almost for four decades. An important breakthrough came in 1983 when de Groot and co-workers discovered an interesting property called half-metallicity with high T_c (deGroot et al., 1984). Research on half-metallic ferromagnetic (HMF) is in progress among the family of Heusler compounds. Kubler and co-workers reported that the majority of Heusler alloys exhibit a semiconducting bandgap at the Fermi energy (E_F) in the minority spin DOS, while the majority spins are metallic. Half-metallic compounds exhibit 100% spin polarization at the Fermi level. The present day technological devices are based on the charge of an electron. For example, in memory devices like dynamic random access memory (DRAM), data are manipulated and stored as charge on capacitors. The disadvantage of charge-based electronic devices is that the information is often lost whenever the power is switched off. This inconvenience can be solved by making size compatible, faster, and reliable devices. An innovation to enslave electron spin is an additional functionality in giant magnetoresistance (GMR) (Baibich et al., 1988; Fert, 2005) in late 1980s, which utilizes the electron spin for information manipulation, storage, and transmission. Spin polarization is a method to define the quantitative degree of half-metallicity; however, a major issue is the definition of the spin polarization P. The electron spin polarization (P) at Fermi energy (E_F) of a material is defined by Soulen et al. (Soulen et al., 1998; Julliere, 1975) as:

$$P = \frac{\rho\uparrow(E_F) - \rho\downarrow(E_F)}{\rho\uparrow(E_F) + \rho\downarrow(E_F)} \tag{9.6}$$

where $\rho\uparrow(E_F)$ and $\rho\downarrow(E_F)$ are the dependent density of states at the E_F. The $\uparrow$ and $\downarrow$ denote the spin-up and spin-down states, respectively.

TABLE 9.1 List of Selected Heusler Alloys (Half-Metals and Semiconductors) Based on Slater–Pauling Rule.

Full-Heusler	a (Å)	Z_t	$M_t = (Z_t - 24)$ (u_B/atom)	Characteristics
1. Co_2MnAl	5.98	28	4.00	Half-metal (FM)
2. Mn_2ZrGe	6.079	25	1.00	Half-metal (FM)
3. Co_2MnSb	5.917	30	6.0	Half-metal (FM)
4. Fe_2Val	5.76	24	0.00	Semiconductor (NM)
5. Fe_2TiSi	5.72	24	0.00	Semiconductor (NM)
6. Mn_2CrSn	6.140	24	0.00	Semiconductor (NM)

TABLE 9.1 *(Continued)*

Half-Heusler	*a* (Å)	Z_t	$M_t = (Z_t - 18)$ (u_B/atom)	Characteristics
1. NiCrSi	5.483	20	2.0	Half-metal (FM)
2. NiCrGe	5.585	20	2.0	Half-metal (FM)
3. CoMnSb	5.87	21	3.0	Half-metal (FM)
4. HfPtPb	6.485	18	0.00	Semiconductor (NM)
5. NiTiSn	5.9534	18	0.00	Semiconductor (NM)
6. TaIrGe	5.966	18	0.00	Semiconductor (NM)

9.4.1.1 Co_2MnAl: A full-Heusler compound exhibiting half-metallic (HM) behavior

After observing several works on X_2YZ series from both theoretical and experimental fronts, it is highly motivating to continue further research to explore for future technological applications. In this section, we have discussed the electronic and magnetic properties of Co_2MnAl, a full-Heusler alloy. Fermi level is used to analyze the electronic properties of the systems with an asymmetry of the band structure. The majority spin band is a metal (i.e., dispersed DOS at E_F), whereas the minority band has semiconducting bandgap (i.e., a bandgap at E_F). The result of metal semiconductor hybrid governs the 100% spin polarization. Therefore, the alloy will have fully spin polarized current and is vital for spin injection in spintronic devices. The presence of small DOS in the conduction band (CB) is due to five majority electrons in the Mn-d orbital. On the other hand, the majority of unfilled states are available in the conduction region mainly for minority electrons. The mechanism that leads to formation of the minority bandgap is a result of d–d hybridization among the d-block elements discussed somewhere else (Galanakis et al., 2006; Webster et al., 2006; Galanakiset al., 2005). It is known that the bandgap in the minority DOS of Co2-based Heusler compounds arises from the hybridization of d electrons of the d-orbitals of Co1 and Co2 atoms and as well as Y atoms sitting adjacent as shown in Figure 9.8. In the first step, the d orbitals of each Co–Co atoms couple as a result of their symmetry that forms bonding (e_g, t_{2g}) and antibonding (e_u, t_{1u}) states. In the second step, the hybridized Co–Co states interact with the d states of Mn atoms. The doubly degenerate Co d-e_g orbitals hybridize with the de_g (d_{z^2}, $d_{x^2-y^2}$) orbitals of the Mn-d states, giving rise to bonding and antibonding states, situated below and above the Fermi level, respectively. The d-t_{2g} Co orbitals couple with the d-t_{2g} (d_{xy}, d_{yx}, d_{zx}) states of the Mn atom, giving rise to a low-lying triplet t_{2g} state with a bonding character and a triplet antibonding t_{2g} above the Fermi level. The remaining antibonding states of Co orbitals (e_u and t_{1u}) do not take part in the hybridization, since there are no Mn atoms with the same symmetry group thus remain non-bonded. Therefore, a real minority gap exists in the Co_2MnAl Heusler compound. The size of the gap is related to the splitting d–d hybridization of Co–Co interaction. In the above discussion, the Al atom has negligible contribution. Al atom helps in positioning the Fermi level within the minority band gap. However, the s and p orbitals have significant role in the distribution of electrons in various distinguishable symmetry states (t_{2g} and e_g) at Co and Mn sites (Kandpal et al., 2006). This will give an importance not only in stabilizing the $L2_1$ structure, but also to predict the magnetic moments at the Co and Mn sites.

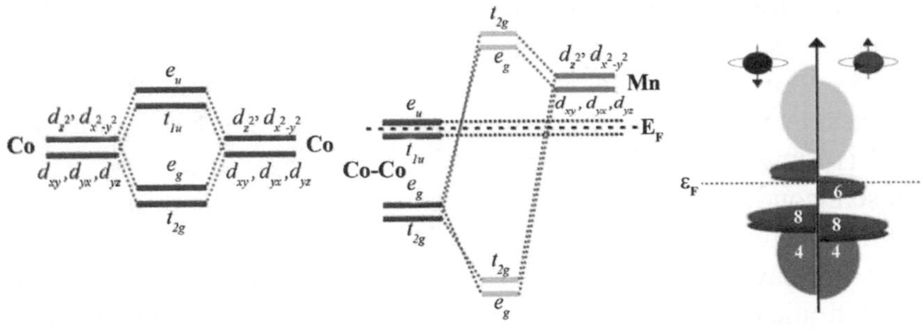

FIGURE 9.8 Schematic illustration of the origin of the minority bandgap in X_2YZ Heusler compounds.

The presence of low-lying s and p states does not contribute directly in the formation of the minority bandgap. These states contribute in counting the total number of elections of occupied and empty states. The energy bands were plotted for Co_2MnAl for both the spin channels as shown in Figure 9.9. For spin up, the Co_2MnAl alloy is a metal in which the majority bands (spin up) crosses the Fermi level (E_F) in rather all higher symmetries. On the other hand, the minority bands (spin down) show a bandgap, the transition is along the highest energies of occupied band at Γ and the lowest unoccupied band lies at the X. This gives an indirect energy bandgap of ~0.70 eV along Δ direction. The partial DOS of each atom are also presented in Figure 9.10. From Figure 9.10 (a–d), it is clear that the effect of Mn-atom is more prominent than Co atom in creating the bandgap in spin down channel.

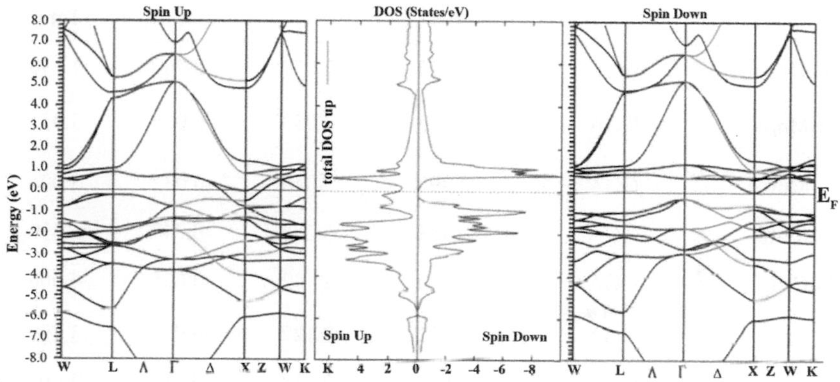

FIGURE 9.9 Half-metallic full-Heusler compound (Co_2MnAl) with spin up band structure (left), total DOS (middle), and spin down band structure (right).

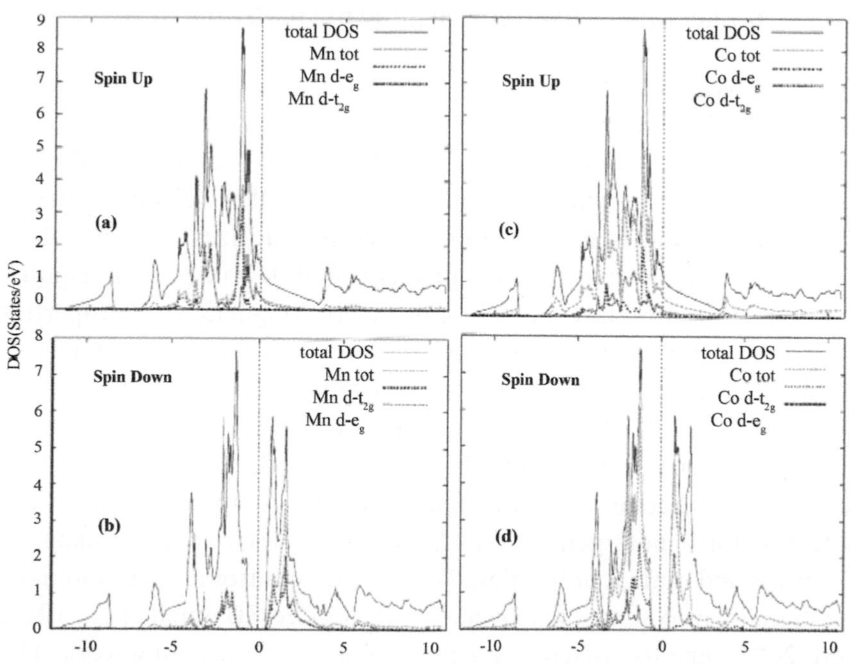

FIGURE 9.10 DOS plots of Co_2MnSn, (a) Co (d, d_{eg}) and Mn (d, d_{eg}) states in spin up; (b) Co (d, d_{eg}) and Mn (d, d_{eg}) states in spin down; (c) Co (d, d_{t2g}) and Mn (d, d_{t2g}) states in spin up and (d) Co (d, d_{t2g}) and Mn (d, d_{t2g}) states in spin down.

TABLE 9.2 Total and Partial Magnetic Moments.

Compounds	XC	Magnetic moments, μ_B			
		Co	Y	Z	Total
Co_2MnAl	GGA	0.740	2.678	-0.060	4.030
Co_2MnSi	GGA	1.029	3.058	-0.055	5.031
Co_2MnGe	GGA	0.991	3.048	-0.032	5.004
Co_2MnSn	GGA	0.963	3.252	-0.047	5.016
Co_2CrSi	GGA	0.980	2.102	-0.055	4.006
Co_2CrGe	GGA	0.932	2.122	-0.029	3.999
Co_2CrSb	GGA	1.058	2.853	-0.014	4.999

9.4.2 *Application of HM Heusler Alloy in Spintronic Devices*

The spin-based electronics that manipulate the electron spin degree of freedom is termed as spintronic, where the spin of an electron is tuned by an applied magnetic field that orients the spin for polarization. These polarized electrons are used to control the electric current. The ultimate goal is to develop a device that utilizes the spin of an electron. Once the spin functionality is added, it will provide significant versatility to future electronic products. Magnetic spin properties of electrons are used in many applications such as giant magnetoresistance (GMR), tunneling magnetoresistance (TMR) (Mathon et al., 2001; Ikeda et al., 2008), magnetic memory (MRAM) (Sbiaa et al., 2011; Bhatti et al., 2017) etc. Materials that undergo phase transition from semiconductor to ferromagnetic above room temperature are potential for a new generation of spintronic devices with enhanced electrical and optical properties. The field of spintronic revolutionized the digital world at post discovery of GMR effect (Baibich et al., 1988; Fert, 2005), and this discovery was the Nobel Prize winning work. The GMR effect occurs due to alignment of the spin of electrons with the applied magnetic field that includes the variation in the resistance of a material. A schematic component of spintronic is presented in Figure 9.11.

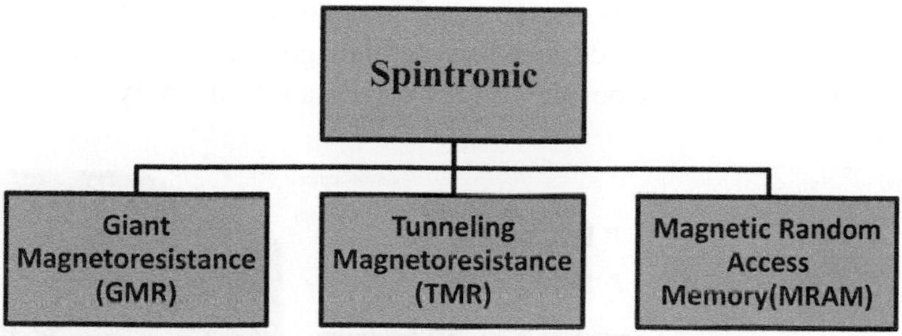

FIGURE 9.11 Spintronic and its components.

GMR is a quantum mechanical effect of magnetoresistance observed in multilayer composites of alternating ferromagnetic and non-magnetic layers as shown in Figure 9.12(a). The impact is to see a noteworthy change in the electrical resistance that relies on the magnetization of the nearby ferromagnetic layers whether they are in a parallel or an antiparallel arrangement. The overall resistance is moderately low for parallel arrangement and generally high for antiparallel arrangement. The direction of magnetic polarization can be controlled by applied magnetic field. The impact relies upon the scattering of electrons on the spin-polarization. TMR is upgradation of spin-valve GMR, in which electrons are incident perpendicular with their spins orientation on layers separated by a thin insulating tunnel barrier (replacing the magnetic layer) as shown in Figure 9.12 (b). This permits to accomplish bigger impedance, a bigger magnetoresistances at negligible temperature. TMR has now outclassed GMR in MRAMs and disk drives, specifically for high surface densities and recording.

The main application of GMR sensor is to read data in hard disk drives, biosensors, micro-electro-mechanical systems (MEMS) (Bhatti et al., 2017) and magneto-resistive random-access memory (MRAM) that stores one bit of information is shown in Figure 9.12 (c). The magnetic memory is based on the storage of spin orientations. The p-type layer depleted in the p-n junction when negative voltage is applied, deforming the spin orientation. This state can be taken as "0" bit or erase memory cell. When the voltage is removed, the concentration of the holes increases. A quantum coupling effect between the holes and the magnetic atom creates the spin alignment. This state can be considered as "1" bit or writes memory cell. The memory devices are equipped with a magnetic sensor etched with layers in the semiconductor. The supersensitive sensor detects the state of magnetization, which exists or not is determined by the device's potentiality as a read memory.

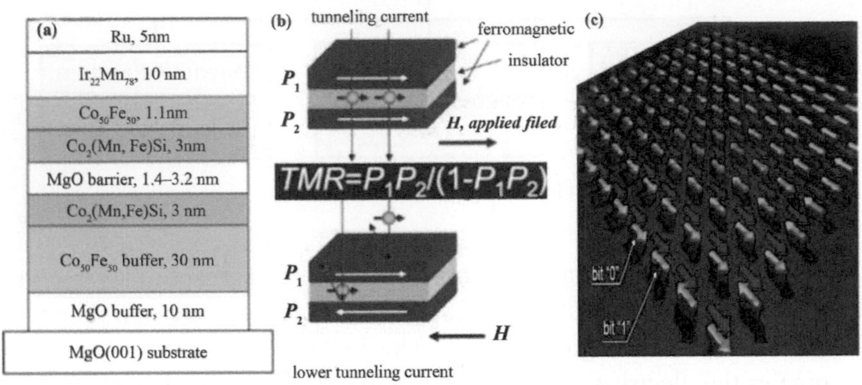

FIGURE 9.12 (a) Schematic diagram of GMR with multilayer structure consisting of ferromagnetic Heusler alloys and semiconducting spacers. (b) Structure of TMR with insulating middle layer between two ferromagnetic Heusler alloys align parallel and antiparallel. (c) Orientation of electron spin in magnetic memory with representation of "0" and "1" bit with respect to electron spin. (Reprinted/adapted a) Liu, et al, 2015. b) Inomata, nd. c) Urbaniak, nd.)

9.5 SEMICONDUCTOR HALF-HEUSLER HFPTPB

HfPtPb is a non-magnetic semiconductor in accordance with the Slater-Pauling rule that follows $M_t = (Z_t - 18)\ \mu_B$/atom. The electronic configuration of HfPtPb is Hf [$5d^2 6s^2$], Pt [$5d^9 6s^1$], and Pb [$6s^2 6p^2$], which offers $Z_t = 18$, as an end result $M_t = 0.0\ \mu_B$/atom. In HfPtPb, the DOS close to the gap are dominated by d-states, in the valence band via bonding hybrids with large Hf or Pt admixture and in the conduction band anti-bonding hybrids with Pt admixture. Thus, the bandgap originates from the strong hybridization between the d-orbitals of the higher valent (Pt) and the lower valent (Hf) transition metallic atoms. This is proven schematically in Figure 9.13 (a).

We can see that the creation of the bandgap in half-Heusler is extremely similar to the bandgap in semiconductors like GaAs that is generated from the hybridization of the low-lying As-sp states with the energetically greater

Ga sp-states. Note that in the half-Heusler (XYZ) with $C1_b$-structure the X and Y sub-lattices form a zinc blende structure, which is essential for the formation of the gap. The distinction with reference to GaAs is only the presence of d-orbitals, i.e. three t_{2g} and two e_g orbitals, involved in the hybridization, as a substitute of sp^3-hybrids in the semiconductors. The measure of bandgap in the half-Heusler generally happens due to the difference between the valence band maxima (VBM) and the conduction band minima (CBM) at the Γ point. For HfPtPb, we obtain a bandgap of about 0.58 eV. As it is well known that the local density approximation (LDA) and the GGA underestimate the values of the gaps in semiconductors as compared to experimental one.

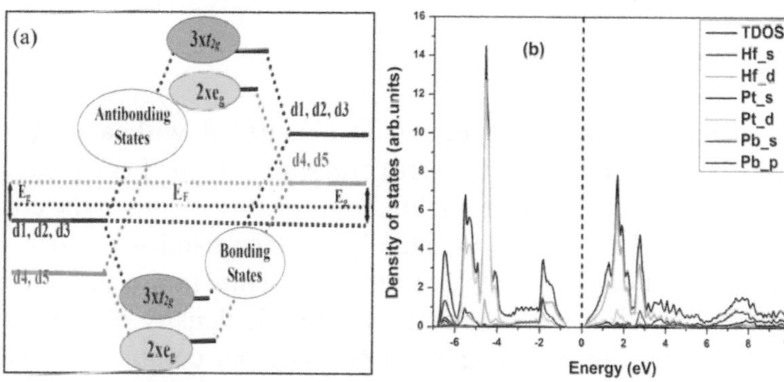

FIGURE 9.13 Schematic illustration of the origin of the minority bandgap in XYZ Heusler compounds and partial DOS of HfPtPb. (Reprinted from Kaur et al., 2017. https://aip.scitation.org/doi/pdf/10.1063/1.4996648).

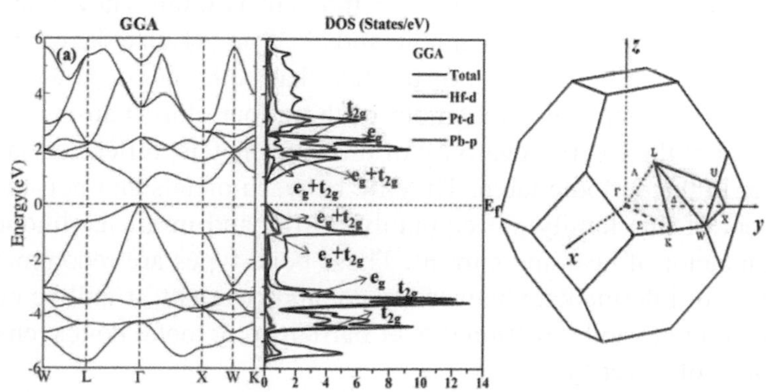

FIGURE 9.14 Electronic band structure and density of states of half-Heusler HfPtPb along with the first Brillouin Zone (Reprinted from Kaur et al., 2017. https://aip.scitation.org/doi/pdf/10.1063/1.4996648).

The calculated band structure and DOS are plotted collectively for better comparison alongside with the first Brillouin area of FCC lattice as depicted in Figure 9.14. As shown in Figure 9.14, the VBM and CBM lie one above the other on excessive symmetry point, which indicates that HfPtPb is a direct bandgap semiconductor. The projected density of states shows that the majority of DOS are due to the dispersed d-states of Hf and Pt close to the Fermi energy, while Pb-p is negligible. The presence of sharp bands in the conduction vicinity suggests that these substances have high energy issue in n-type composition. Thus, we can easily articulate that d-electrons play a most important role in thermoelectric properties.

9.6 SEMI-METAL

A semi-metallic Heusler compound is a material with a very small quantity of band (conduction/valence) crossing over the Fermi energy, or there may additionally be overlapping between the bottom of the conduction band and the top of the valence band. One band is almost filled, whereas the other one is almost empty at zero temperature. A semi-metal for that reason has no bandgap and a negligible density of states at the Fermi level. They are the category of two Heusler compounds with the presence of total electron count $Z_t = 24$ (full-Heusler) or $Z_t = 18$ (half-Heusler) near Fermi level and strictly follow Slater-Pauling rule of $M_t = (Z_t - 24)$ μB/atom or $M_t = (Z_t - 18)$ μB/atom relying upon their composition, which indicates the magnetic moment is 0.00. They also have small effective masses for each hole and electron because the overlap in strength is typically the result of the fact that both bands are wide. The covalent bonds becoming a member of these four closest atoms may be interfaced of the electronegativity of the semi-metals, reflected via their position in the periodic table. They lie between metals and non-metals. Members of this family are equitably brittle, and in all likelihood are bad conductor of heat and current. These bond types are recommended by means of intermediate between metal and covalent; it will be consequently stronger and directional over participating metal bonds, ensuing in crystals of lower symmetry.

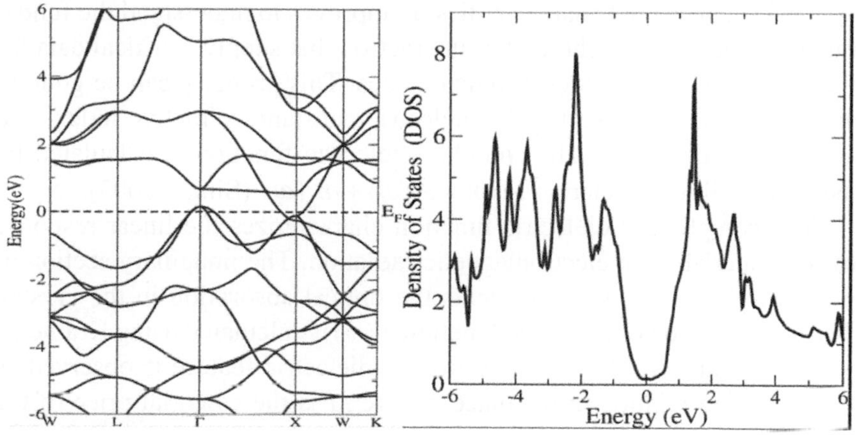

FIGURE 9.15 Semi-metallic DOS and band structure of Ru_2TaAl.

As semi-metals have lower charge carriers than metals, they typically have low electrical and thermal conductivities. In addition, they usually show high diamagnetic susceptibilities and lattice dielectric constants. Some of the examples of recently discovered semi-metallic compounds among the classes of half-Heusler and full-Heusler alloys are NdPtBi, YPtBi, ScPtBi, GdPtB, LiBaBi, HoPdBi, YAuPb, LuPdSb, CrTiVAl, Ru_2TaAl, Ru_2NbGa, Ru_2CrIn, Fe_2VGa, and more to explore. Ru_2TAl is a type of semi-metallic Heusler alloy whose energy band and density of states are presented in Figure 9.15. First principle calculations revealed an indirect overlap between electron and hole pockets that gives a pseudogap in the neighborhood of E_F (Angell et al., 1983; Tseng et al., 2017). Such a state of affairs provides a practical interpretation for the semi-metallic behavior in Ru_2TaAl, as shown in Figure 9.15.

9.7 OPTICAL PROPERTIES OF SEMICONDUCTOR HEUSLER ALLOY

The study about the optical behavior of a solid is related with the digital structure of a material. This is due to the reality that an incident light (photon energy) relies on its electromagnetic area engage with the nearby electromagnetic field of atoms usually offers optical response, as shown in Figure 9.16(a and b). As a result, the mild that emerges out may also no longer have the equal characteristics. It approves to understand the fundamental principles of light–matter interaction that supply a critical data for the development of optoelectronic devices. This property can be good to analyze through calculating the dielectric constant, refractive index, and absorption coefficient. The complex dielectric function is calculated by using the following relation $\varepsilon(\omega) = \varepsilon_1(\omega) + i\varepsilon_2(\omega)$ (Singh, 2003).

The complicated dielectric function characterizes the linear response of the material to an electromagnetic radiation. The imaginary section of dielectric function $\varepsilon_2(\omega)$ represents the optical absorption in the crystal. The actual section of dielectric function $\varepsilon_1(\omega)$ is calculated using Kramers–Kroning relations. The actual section of dielectric feature is observed as the following relation in the place, where M is the essential price of the integral.

$$\varepsilon_1(\omega) = 1 + \frac{2}{\pi} M \int_0^{\infty} \frac{\varepsilon_2(\omega')\omega'}{\omega'^2 - \omega^2} d\omega \tag{9.7}$$

The imaginary part of the dielectric function $\varepsilon_2(\omega)$ can be calculated by using the momentum matrix elements between the occupied and unoccupied wave functions as following relation:

$$\varepsilon_2(\omega) = \frac{Ve^2}{2\pi\hbar m^2\omega^2} \int d^3k \sum_{n,n'} \left| \langle kn | p | kn' \rangle \right|^2 f(kn)(1 - f(kn'))\delta(E_{kn} - E_{kn'} - \hbar\omega) \tag{9.8}$$

where V is the volume of the unit cell; e is the electron charge, m is its mass, two is the wave feature of the crystal; $|kn\rangle$ is the wave function of the crystal; $f(kn)$ is the Fermi-Dirac two-function distribution; E is the incident photon energy; n refers the unoccupied wave features; and n two refers the occupied wave functions.

The incident photons on the semiconductor offer response to such exterior mild source or the light generated by means of the semiconductor below an excitation source. Both phenomena are extremely touchy to the electronic band-gap. The imperative band hole is the threshold electricity for one-of-a-kind optical phenomena to be observed. In fact, the semiconductor is obvious to light with strength under the bandgap, whilst it turns into opaque to mild with strength above the bandgap. The optical response is collected in the structure of spectral traces in affiliation with the electron transition between the two allowed energy Eigen states, as shown in Figure 9.16 (a and b). In most of the semiconductor half-Heusler alloys, the band hole tiers from 0.20 to 1.00 eV. This indicates that HH semiconductors are optically lively underneath IR–UV (Infrared-Ultraviolet) range.

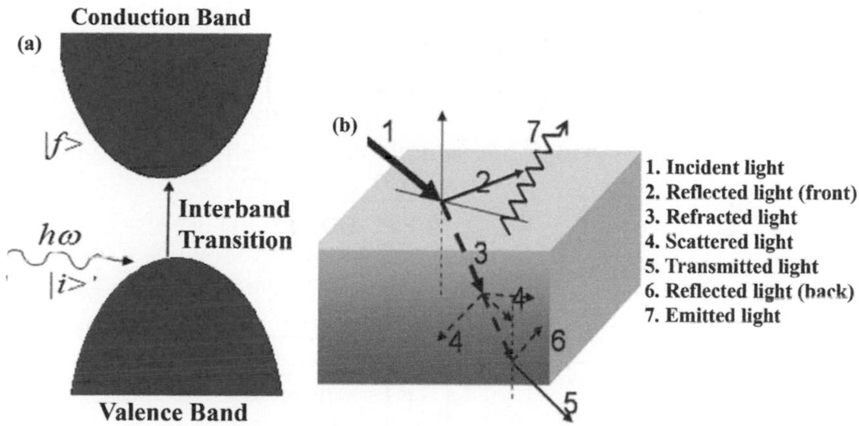

FIGURE 9.16 (a) Interband transition of an electron from initial state (k_i) to final state (k_f) due to the interaction of photon ($\hbar\omega$) with atoms. (b) Scheme of the different processes of interaction between light and a solid. 1: incident light, 2: reflected light (front surface), 3: refracted light, 4: scattered light, 5: transmitted light, 6: reflected light (back surface), 7: emitted light (Adapted from Singh, 2003).

Figure 9.17 illustrates the calculated dielectric characteristic of the half-Heusler XYZ kind semiconductor half-Heusler compound. It can be considered that the imaginary spectrum of $\varepsilon(\omega)$ has an outstanding absorption peak, positioned at the photon energy ~2.5 eV. A excessive intensity $\varepsilon_2(\omega)$ is discovered in a vary between 0.2 and 4 eV, which indicates the absorption of photon energy by using atoms. The expressions of other optical residences like refractive index $n(\omega)$, extinction coefficient $k(\omega)$, reflectivity $R(\omega)$, and energy loss function $L(\omega)$ are calculated through the usage of dielectric features and represented through eq 9.9 (Shen et al., 2008; Dadsetani et al., 2006).

$$n(\omega) = \left(\frac{\varepsilon_1(\omega)}{2} + \frac{\sqrt{\varepsilon_1^{\,2}(\omega) + \varepsilon_2^{\,2}(\omega)}}{2} \right)^{1/2} \tag{9.9}$$

$$k(\omega)=\left(\frac{\sqrt{\varepsilon_1^{\,2}(\omega)+\varepsilon_2^{\,2}(\omega)}-\varepsilon_1(\omega)}{2}\right)^{1/2} \tag{9.10}$$

$$R(\omega)=\left(\frac{(n(\omega)-1)^2+k(\omega)^2}{(n(\omega)+1)^2+k(\omega)^2}\right) \tag{9.11}$$

$$L(\omega)=\left(\frac{\varepsilon_2(\omega)}{\varepsilon_1^{\,2}(\omega)+\varepsilon_2^{\,2}(\omega)}\right) \tag{9.12}$$

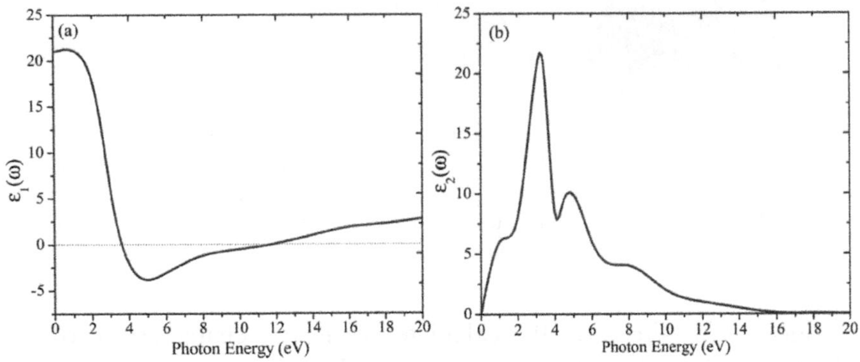

FIGURE 9.17 (a) Real part of dielectric function ($\varepsilon_1(\omega)$). (b) Imaginary part of dielectric function $\varepsilon_2(\omega)$ of semiconductor half-Heusler compound.

Using these relations, the determined optical parameters are plotted in Figure 9.18 (a–d). According to the dispersion curve of refractive index at $\omega = 0$, $n(\omega)$ is found to be 4.6 and the extinction coefficient $k(\omega)$ is estimated as 1.5 as shown in Figure 9.18 (a and b). The higher value of static refractive index proves material's interaction with the light. Reflectivity has less peak values in contrast to loss function. The energy loss functions $L(\omega)$ describe the energy loss when an electron passes through a medium. The main peak in the energy loss functions $L(\omega)$ is located at ~12.5 eV, due to the occurrence of plasmon frequency (ω_p).

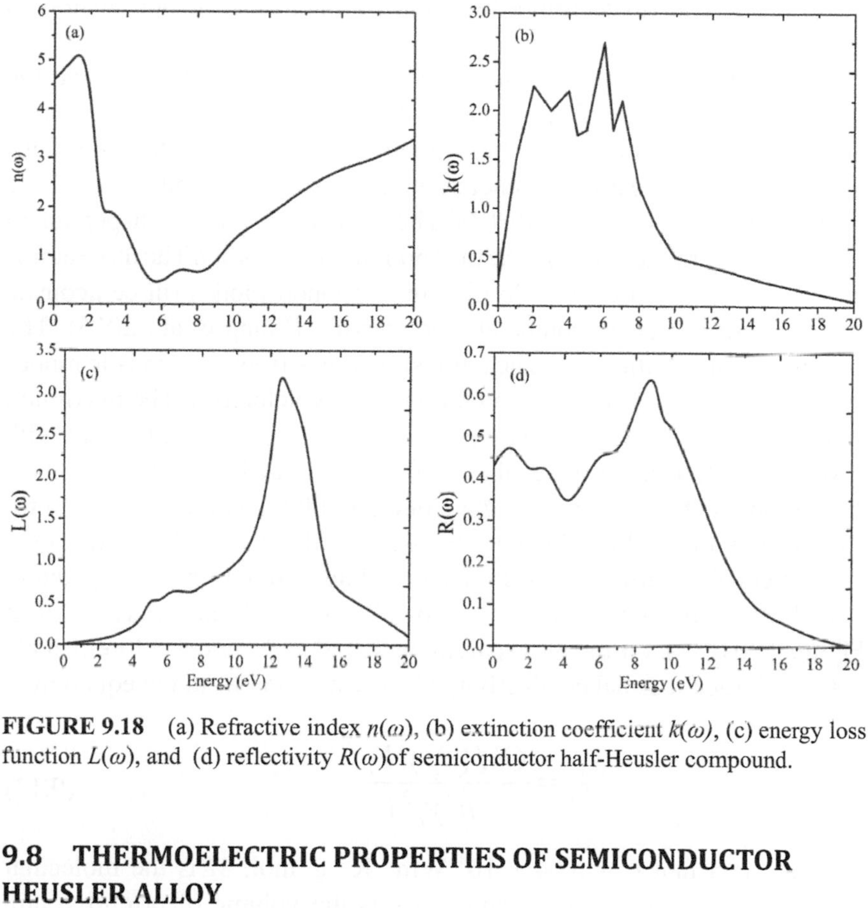

FIGURE 9.18 (a) Refractive index $n(\omega)$, (b) extinction coefficient $k(\omega)$, (c) energy loss function $L(\omega)$, and (d) reflectivity $R(\omega)$of semiconductor half-Heusler compound.

9.8 THERMOELECTRIC PROPERTIES OF SEMICONDUCTOR HEUSLER ALLOY

Since the industrial application of a product is constantly a matter of the price, one important assignment for the on-going thermoelectric materials is to discover the reasonably priced elements necessary for making XYZ compound while preserving an excessive discern of merit ZT. HH compounds with semiconducting or semi-metallic HHs are the novel candidates for thermoelectric materials due to their splendid mechanical and thermal stability. The distinctiveness being the presence of factors having high melting point such as Hf: 2231 °C, Zr: 1855 °C, Ti: 1668 °C, Ni: 1455 °C, and Co: 1495 °C as properly as elements with low melting factor such as Sn: 232 °C and Sb: 631 °C. The calculation of phonon dispersion is a key to determine the thermodynamical structural steadiness of a system.

Figure 9.19 indicates the calculated phonon dispersion $hv(q)$ and the accompanied density of states $g(\omega)$ of HfPtPb. The topmost, separate bands occur in the main from the optical modes related to the vibrations of the Pb atoms. The high-symmetry phonon dispersion band occurs along the Γ-point, as shown in Figure 9.19. The presence of phonon dispersion in real frequency range confirms the thermodynamic stability. The three atoms in the primitive cell of HfPtPb give nine phonon modes (three acoustic and six optical modes) comparable to HfNiPb (Wang et al., 2016). The maximum contribution to the heat transfer comes from the acoustic modes due to their robust dispersion and massive crew velocities. The maximum amplitude of acoustic frequency is ~100 cm^{-1}, in precise settlement with 120 cm^{-1} of HfNiPb (Wang et al., 2016). The mixing between the acoustic and the optical branches gives a giant phonon–phonon scattering, accountable for low thermal conductivity (Gudelli et al., 2015). Thermodynamic parameters like Grüneisen parameter, volume, and Debye temperature have been evaluated from Gibbs (Roza et al., 2011) and introduced in Figure 9.20 (a–c). Fitting these parameters in Slack's model (Morelli et al., 2006), a lattice thermal conductivity (k_l) is estimated from the equation:

$$k_l = \frac{CM(\theta_D)^3 V_T^{\ 1/3}}{n^2 \gamma_T^{\ 2} T} \tag{9.13}$$

Here, constant $C = 3.04 \times 10^4$ $Wm^{-2} K^{-1} g^{-1}mol$; M is the molecular mass; Θ_D is the Debye temperature; V_T is the volume of primitive unit cell; n is the number of atoms in unit cell; γ_T is Gruneisen parameter; and T is the absolute temperature. For validation of the results obtained from Slack's equation, and ShengBTE (Li et al., 2014), the results of thermal conductivities are presented together in Figure 9.20 (d). The room temperature values of k_l from Slack's equation and ShengBTE code are found to be in close approximation, ~7.1 and 9.1 W mK^{-1}, respectively. The material performance is in line with the published values for comparable material compositions and exhibits high ZT-values ~0.7 for n-type Heusler compound close to a bench mark value of ZT = ~1 n-type and 0.5 for the p-type samples (Kaur et al., 2017). The modules constructed from these materials have a maximum electricity output of 2.8 W, and the electricity density of 3.2 W cm^{-2} and a Z-value of 3.1×10^{-4} K^{-1} for

HH modules published so far (Kaur et al., 2017). Effects of the phonon dispersion are based on the electronic structure calculations and harmonic force constants as acquired in Quantum Espresso package (Giannozzi et al., 2009).

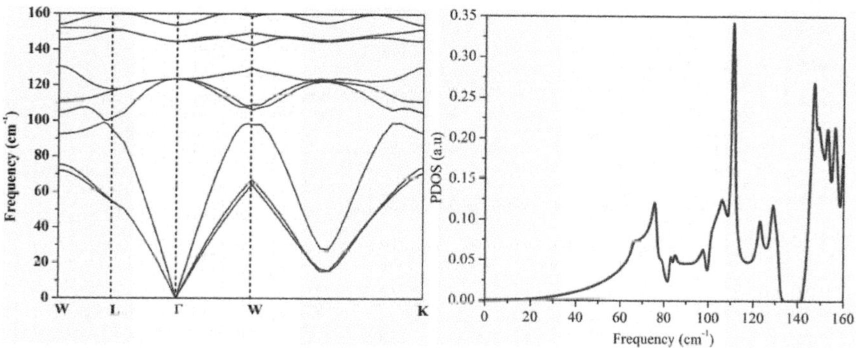

FIGURE 9.19 Phonon dispersion (band and DOS). (Reprinted from Kaur et al., 2017. https://aip.scitation.org/doi/pdf/10.1063/1.4996648)

A thermoelectric property of a crystalline stable is in close relation with its electronic band energy. The presence of narrow bandgap (0.2–1.8) eV and sharp DOS near E_F is beneficial for high Seebeck coefficient and excessive electrical conductivity. Thus, the thermoelectric parameters like Seebeck coefficient (S), electrical conductivity per relaxation time (σ/τ), and thermal conductivity per relaxation time (κ/τ) were calculated from the semi-classical Boltzmann transport equation in mixture with the first idea's band-structure within a rigid-band approximation (RTA) as carried out in BoltzTraP (Madsen et al., 2006). A presence of flat band in electronic band shape is due to the large effective mass that leads to enhanced Seebeck coefficient. On the contrary, the sharp band corresponds to low effective mass that is responsible for excessive cost of electrical conductivity. The calculated electron thermoelectric parameters as a function of temperature are in Figure 9.21 (a–d). A thermoelectric efficiency of a material can be measured by using a dimensionless $ZT = \frac{S^2 \sigma T}{\kappa_l + \kappa_e}$, where κ_e and κ_l are the thermal conductivities due to electron and lattice contribution, respectively.

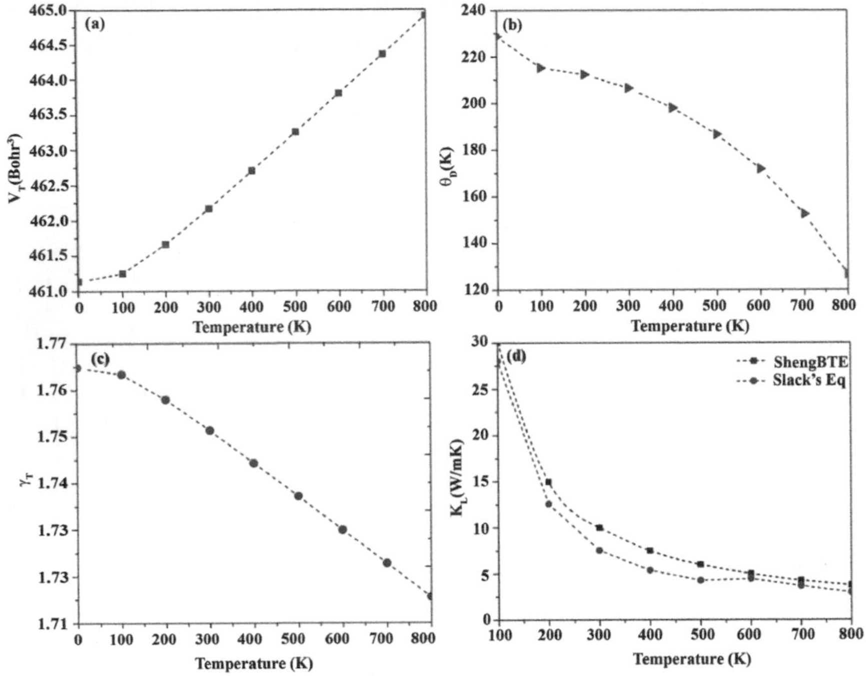

FIGURE 9.20 Thermodynamical parameters: (a) volume, (b) Debye temperature, (c) Gruneisen parameter, and (d) lattice thermal conductivity (from Slack's equation and ShengBTE) as a function of temperature. (Reprinted from Kaur et al., 2017. https://aip.scitation.org/doi/pdf/10.1063/1.4996648).

The *ZT* value above ~1.0 is the benchmark for realistic application of a material. The calculated thermoelectric efficiency of HfPtPb has been compared with the theoretical result of analogous compounds (HfNiPb and HfPdPb). In Figure 9.21 (a), we found the magnitude of *S* increases with growing temperature in negative axis indicating n-type charge carriers. The absolute value of *S* is -95 μV K^{-1}, in shut agreement to -100 μV K^{-1} of HfNiPb and HfPdPb at 300 K (Pénicaud et al., 1992). In Figure 9.21 (b), the electrical conductivity (σ/τ) shows the decreasing trend with increase in temperature. The room temperature value of σ/τ is 12.4 × 10^{18} (Ω ms)$^{-1}$. The direct comparison of σ/τ with the experimental σ or resistivity is not justified due to an unknown constant relaxation time (τ). Also, the measurement of total thermal conductivity ($k_e + k_l$) is not possible; hence k_e/τ and k_l are presented independently in Figure 9.21 (b and c). Knowing all the electron transport parameters, a dimensionless figure of merit ZT_e is calculated and presented in Figure 9.21 (d). The room temperature value of ZT_e is ~0.32, which is too low for energy conversion at intermediate temperature range. However, the ZT_e is continuously increases linearly with temperature reaching a value as high as 0.65 at high temperature indicating that the material is suitable for high temperature thermoelectric energy conversion.

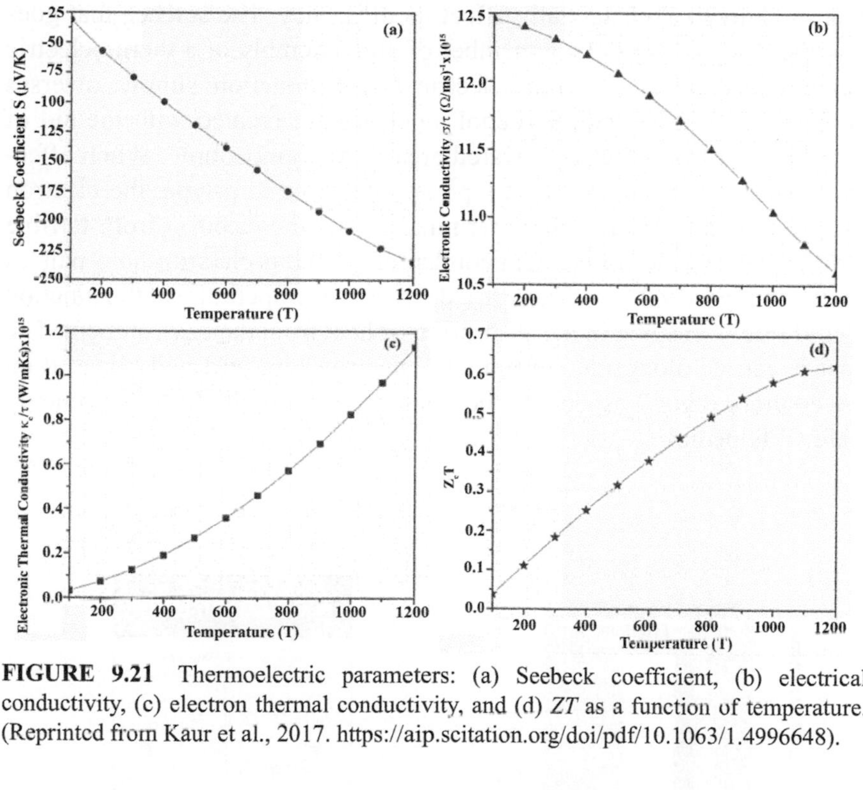

FIGURE 9.21 Thermoelectric parameters: (a) Seebeck coefficient, (b) electrical conductivity, (c) electron thermal conductivity, and (d) *ZT* as a function of temperature. (Reprinted from Kaur et al., 2017. https://aip.scitation.org/doi/pdf/10.1063/1.4996648).

9.9 DESIGNING AND WORKING OF SOLID-STATE THERMOELECTRIC MATERIAL

A schematic module for thermoelectric generator (TEG) and cooling is presented in Figure 9.22 (a and b). A thermoelectric module consists of two dissimilar solid thermoelectric materials joining their ends: an n-type and p-type semiconductors that form a couple. A TEG module operates for a very large temperature gradient. A thermocouple will absorb heat at the heat source, and the respective charge carriers move downward toward the heat sink, as a result negative and positive charges accumulate at the end of n-type and p-type terminals, respectively, as shown in Figure 9.21 (a).

Now, there exists a positive and negative potential, when connected with a wire a circuit will form and a direct electric current will flow. This is to make certain that a continuous circuit can be made whereby current can float and electricity can be produced. Furthermore, the geometrical graph of a module will substantially affect its efficiency. The science that goes into the design, becoming a member of and assembly of a thermoelectric module is copious. A thermoelectric cooler function simply reverses to the TEG. The warmth and cool junctions are created with the aid of making use of the electric powered area in a thermocouple. When effective terminal is connected to n-type and negative to p-type, the electron from n-type diffuses to superb terminal, while the electron from terrible terminal enters the p-type semiconductor growing electron-hole pair as shown in Figure 9.21 (b). There is a decrease in temperature at the junction ("cold side"), ensuing in the absorption of heat from the environment. The heat is carried alongside elements by electron transport and released on the contrary ("hot") aspect as electrons cross from a high- to low-energy state (Wikipedia).

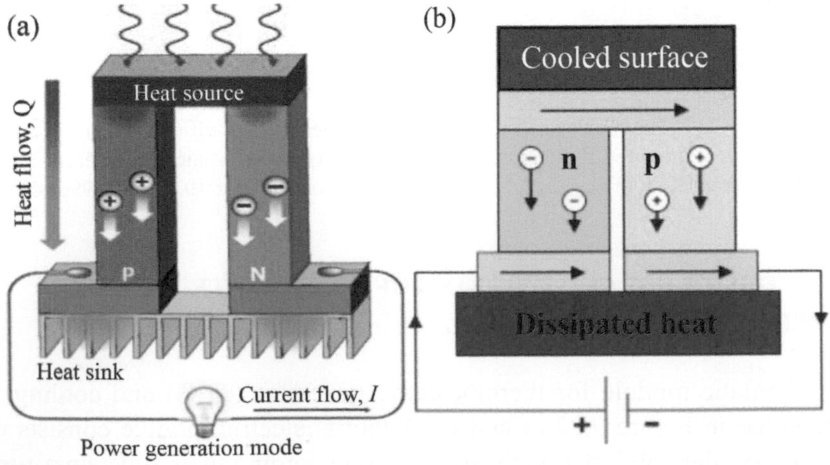

FIGURE 9.22 A schematic thermoelectric module of (a) electricity generator and (b) cooling.

9.10 CONCLUSIONS AND SUMMARY

This portion summarizes the chapter starting from electronic to thermoelectric properties. The Heusler compounds are a charming category of future materials, exhibiting various novel properties that can be utilized in a vast large spectrum of applications. Herein, we have mentioned the formation of electronic bandgap and magnetic properties of half-metallic Heusler alloys, which signify fascinating hybrids between metal and semiconductor. The presence of bandgap in the minority band is a fascinating feature, and consequently the appreciation of the gap is very crucial. For instance, Co_2MnAl is a half-metal ferromagnet with a bandgap at the minority spin and has an integer total spin moment of $M_t = Z_t - 24 = 4.0\ \mu_B$/atom, which is an end result of the wide variety of uncompensated spins. In addition to the Co–Co bonding and antibonding d-hybrids, Co states cannot hybridize with the Mn and the Al atoms immediately and are exclusively localized at the two Co sub-lattices. Thus, in addition to the five Co-Mn bonding and five Co-Mn antibonding bands, there exist five such non-bonding bands that are solely split-up by the weaker Co–Co hybridization into three occupied d-states of t_{1u} symmetry and two unoccupied e_u states, which are located simply beneath and just above the E_F such that the indirect gap in these materials is smaller than in the half-Heusler. However, the system like Fe_2VAl with $Z_t = 24$ valence electrons are semiconductors; Co_2VAl (26 valence electrons) has a whole spin moment of 2 μ_B/atom; Co_2CrAl 3.0 μ_B/atom, Co_2MnAl 4 μ_B/atom, and Co_2MnSi, which have 29 valence electrons, have a total spin second of 5 μ_B/atom.

The presence of integer value of complete magnetic moment and the two minority band holes at the Fermi degree offers a 100% spin polarization. The 100% spin polarization at E_F is a finger print of half-metal ferromagnets. Half-metals are believed to have a notable potential in spintronics. For the half-Heusler alloys like HfPtPb, which crystallized in $C1_b$ structure, the gap arises from the hybridization between the d-wave features of the lower valent transition metallic atom (Hf) and the d-wave features of the higher valent transition metallic atom (e.g., Pt). Thus, the d–d hybridization leads to 5-occupied bonding bands, which have a large Pt and smaller Hf admixture. These states shape the valence band, being separated with the aid of a band hole from the conduction band, which is formed by using five antibonding hybrids with a large Hf-d and a small Pt-d admixture. The roles of the spatoms like Pb are very important for the bonding, especially for the stabilization of the $C1_b$ structure. The sp-states are created for every spin direction one s and three p bands in the power vicinity beneath the d states which with the aid of hybridization can also accommodate transition metal electrons. In this way, the fine number of valence d-electrons can be changed via the valence of the sp elements.

The minority valence band consists of 9 bands comprising 18 valence electrons and has the equal density of states for both spin directions. More general, compound with a complete quantity of valence electrons $Z_t = 18$ per unit cell (HfPtPb) will provide zero magnetic moment in accordance to Slater–Pauling Rule. Thus, HfPtPb is a non-magnetic semiconductor having direct bandgap ~0.58 eV. The semiconductor half-Heusler offers an appropriate optical response due to the bandgap matching with the IR–UV range. The chemical and physical properties of a stable material are in close relation with the electronic band energy. Hence to get a correct description about a specific fabric its electronic band structure has to be determined accurately. The optical properties of semiconductor half-Heusler compound is explored with the aid of calculating the real-imaginary part of dielectric function, absorption coefficient, electron energy loss function, and refractive index. The first distinct absorption spectra as a function of photon energy rise at ~0.40 eV and a type of robust peak at ~2.0 eV. This shows that HH semiconductor is optically live underneath IR–UV range. A linear increase of absorption peak with growing photon energy was observed, indicating a conical form of the electronic structure in the neighborhood of the Fermi energy. Peaks in the imaginary section of dielectric are regularly resulting from interband transitions. The static refractive index at $\omega = 0$, $n(\omega)$ is found to be 4.6 and the extinction coefficient $k(\omega)$ is estimated as 1.5. The higher cost of static refractive index proves two material's interaction with the light. A prominent height of two-electron loss characteristic and $L(\omega)$ at photon strength 12.5 eV are the occurrence of a two-plasmon frequency (ω_p).The concept of energy conversion, particularly the conversion of the waste heat generated from heavy industries into useable form of energy from a solid state material is a new innovation. One can say that the HH compounds fulfill most of the industrial want for TE materials, i.e. environment-friendly, price effective, durable, structural stability, processable in industrial production, and chemical and mechanical resistance toward excessive temperatures. A thermodynamical stability of HfPtPb is validated by calculating the phonon dispersion. A theoretical calculation predicts that HfPtPb is a viable novel thermoelectric material with low value of lattice thermal conductivity, ~7.5 W mK^{-1} at room temperature. The linear value of the *ZT* value is an evidence for its applicability at greater temperatures.

KEYWORDS

- **Heusler compound**
- **spintronic**
- **thermoelectric**
- **optoelectronic**
- **semi-metal**
- **full-Heusler alloy**
- **half-Heusler alloy**

REFERENCES

Angell, B. J.; Terry, S. C.; Barth, P. W. Silicon micro-mechanical devices. *Sci. Am.* **1983,** *248* (4), 44–55.

Baibich, M. N.; Brote, J. M.; Fert, A.; Nyugen, V. D. N.; Petroff, F.; Eitenne, P.; Creuzet, G.; Friederich, A. Chazelas, J. Giant Magnetoresistance of (001)Fe/(001)Cr Magnetic Superlattices. *Phys. Rev. Lett.* **1988,** *61*, 2472.

Bhatti, S. et al. Spintronics based random access memory: a review. *Mater. Today* **2017,** *20*, 530–548.

Blaha, P.; Schwarz, K.; Madsen, G. K. H.; Kvasnicka, D.; Luitz, J. WIEN2k, An augmented plane wave + Local Orbitals Program for calculating Crystal Properties, Schwarz K., Techn. Universitat Wien, Wien, Austria, 2001, ISBN 3-9501031-1-2.

Bowen, M.; Bibes, M.; Barthélémy, A.; Contour, J. P.; Anane, A.; Lemaître, Y.; Fert, A.; Are half-metallic ferromagnets half metals. *Appl. Phys. Lett.* **2003,** *82*, 233.

Bradley, A. J.; Rogers, J. W. The Crystal Structure of the Heusler Alloys. *Proc. Roy. Soc.* **1934,** *A144*, 340359.

Casper, F.; Graf, T.; Chadov, S.; Balke, B.; Felser, C. Half-Heusler compounds: novel materials for energy and spintronic applications. *Semicond. Sci. Technol.* **2012,** 27, 063001.

Castelliz, L. Studies on the heusler alloys—III. The antiferro-magnetic phase in the Cu-Mn-Sb system. *Mh Chem* **1952,** *82*, 1314.

Chadov, S.; Kuebler, J. X. Qi.; Fecher, G. H.; Felser, C.; Zhang. S. C. Tunable multifunctional topological insulators in ternary Heusler compounds. *Nat. Mater.* **2010,** 9, 541.

Clark, S. J.; Segall, M. D.; Pickard, C. J.; Hasnip, P. J.; Probert, M. I. J.; Refson, K.; Payne, M. C. First Principles Methods Using CASTEP. *Zeitschrift für Kristallographie* **2005,** *220*, 567–570. DOI:10.1524/zkri.220.5.567.65075

Dadsetani, M.; Pourghazi, A. Optical properties of strontium mono-chalcogenides from first principles. *Phys. Rev. B* **2006,** *73*, 195102.

De Groot, R. A.; Mueller, F. M.; van Engen, P. G.; Buschow, K. H. J. New Class of Materials: Half-Metallic Ferromagnets. *Phys. Rev. Lett.* **1984,** *50*, 2024.

Dewhurst, J. K.; Sharma, S.; Nordstrom, L.; Cricchio, F.; Granas, O.; E. K. U. Gross The Elk Code Manual Version 2011, ver. 4.3.6

Fecher, H. G.; Kandpal, H. C.; Wurmehl, S.; Morais, J.; Lin, H. J.; Elmers, H. J.; Schonhense, G.; Felser, C. Superconductivity in the Heusler Family of Intermetallics. *J. Phys.: Cond. Mat.* **2005,** *17*(46), 7237–7252.

Fert, A.; Symposium B. Multi-component Alloys and Intermetallic Compounds for Magnetic Applications and Nanotechnology and E-MRS **2005**, Symposium D, Magnetoelectronics, E-MRS.

Galanakis, I.; Dederichs, P. H. *Half-metallic Heusler alloys: Fundamentals and applications, Lecture Notes in Physics*. Springer: Berlin, 2005; *676*.

Galanakis, I.; Mavropoulos, P.; Dederichs, P. H. The influence of spin–orbit coupling on the band gap of Heusler alloys. *J. Phys. D: Appl. Phys.* **2006,** *39*, 765.

Giannozzi, P. et al. Interatomic potentials for modelling radiation defects and dislocations in Tungsten. *J. Phys. Condens. Mat.* **2009,** *21*, 395502.

Gonze, X. et al. First-principles computation of material properties: the ABINIT software project. *Comp. Mat. Sci.* **2002,** *25*, 478.

Gudelli, V. K.; Kanchana, V.; Vaitheeswaran, G.; Singh, D. J.; Svane, A.; Christensen, N.E. Electronic structure, transport, and phonons of SrAgChF (Ch = S, Se, Te): Bulk superlattice thermoelectrics. *Phys. Rev. B* **2015,** *92*, 045206.

Gulans, A.; Kontur, S.; Meisenbichler, C.; Nabok, D.; Pavone, P.; Rigamonti, S.; Sagmeister, S.; Werner, U.; Draxl, C. Exciting — a full-potential all-electron package implementing density-functional theory and many-body perturbation theory. *J. Phys.: Condens. Matter* **2014,** *26*, 363202.

Heusler, F. ÜbermagnetischeManganlegierungen. *Verhandlungen der Deutschen Physikalischen Gesellschaft (in German)* **1903,** *12*, 219.

Hohenberg, P.; Kohn W. Inhomogeneous Electron Gas. *Phys. Rev. B* **1964,** *136*, 864–871.

Hirohata, A. et al. Heusler alloy/semiconductor hybrid structures. *Curr. Opin. Solid State Mater. Sci.* **2006,** *10*, 93–107.

Ikeda, S.; Hayakawa, J.; Ashizawa, Y.; Lee, Y. M.; Miura, K.; Hasegawa, H.; Tsunoda, M.; Matsukura, F.; Ohno, H. Tunnel magnetoresistance of 604% at 300K by suppression of Ta diffusion in CoFeB/MgO/CoFeBCoFeB/MgO/CoFeB pseudo-spin-valves annealed at high temperature. *Appl. Phys. Lett.* **2008,** *93*(8), 082508.

Inomata, K. nd. Tunnel Magnetoresistance (TMR). Research Center for Magnetic and Spintronic Materials (CMSM), National Institute for Materials Science (NIMS). https://www.nims.go.jp/mmu/tutorials/TMR.html

Ishikawa, M., et al. *Heusler alloys, in Superconductivity in d- and f-Band Metals, Landolt-Börnstein-Group III Condensed Matter*; Buckel, W., Weber W., Ed.; Springer: Kernforschungszentrum, Karlsruhe, 1982.

José S. M., et al. The SIESTA method for ab initio order-Nmaterials simulation. *J. Phys.: Condens. Mat.* **2002,** *14* (11), 2745–2779.

Julliere, M. Tunneling between ferromagnetic films. *Phys. Lett. A* **1975**, *54*, 225.

Kandpal, H. C.; Fecher, G. H.; Felser, C.; Schönhense, G. Correlation in the transition-metal-based Heusler compounds Co_2MnSi and Co_2FeSi. *Phys. Rev. B* **2006**, *73*, 094422.

Kaufmann, S.; Rößler, U.; Heczko, O.; Wuttig, M.; Buschbeck, J.; Schultz, L.; Fähler, S. Adaptive Modulations of Martensites. *Phys. Rev. Lett.* **2016,** *104*, 145702.

Kaur, K.; Rai, D. P.; Srivastava, S. Structural, electronic, mechanical, and thermoelectric properties of a novel half-Heusler compound HfPtPb. *J. Appl. Phys.* **2017,** *122*, 045110.

Knowlton, A. A.; Clifford, O. C. The Heusler alloys. *Trans. Faraday Soc.* **1912,** *8*, 195–206.

Kohn, W.; Sham, L. J. Self-Consistent Equations Including Exchange and Correlation Effects. *Phys. Rev. A* **1965,** *140*, 1133–1138.

Kresse, G. VASP Group, Theoretical Physics Departments, Vienna, March 31, 2010.

Li, W.; Carrete, J.; Katcho, N. A.; Mingo, N. First-principles study on lattice thermal conductivity of thermoelectrics HgTe in different phases. *Comput. Phys. Commun.* **2014,** *185*, 1747.

Lin, H.; Wray, L. A.; Xia, Y.; Xu. S.; Jia, S.; Cava, R. J.; Bansil, A.; Hasan, M. Z. Half-Heusler Ternary compounds as new multifunctional experimental platforms for topological quantum phenomena. *Nat. Mater.* **2010,** *9*, 546.

Liu, H.; Kawami, T., Moges, K.; Uemura, U.; Yamamoto, M.; Shi, F.; Voyles, P. (2015). Influence of film composition in quaternary Heusler alloy Co 2 (Mn,Fe)Si thin films on tunnelling magnetoresistance of Co 2 (Mn,Fe)Si/MgO-based magnetic tunnel junctions. Journal of Physics D: Applied Physics. 48. 164001. 10.1088/0022-3727/48/16/164001.

Madsen, G. K. H., et al. BoltzTraP. A code for calculating band-structure dependent quantities. *Comput. Phys. Commun.* **2006,** *175*, 67.

Mathon, J.; Umerski, A. Theory of tunneling magnetoresistance of an epitaxial Fe/MgO/Fe(001) junction. *Phys. Rev. B*, **2001,** *63* (22), 220403.

Morelli, D. T.; Slack, G. A. High Lattice Thermal Conductivity Solids. In *High Thermal Conductivity Materials*; Shinde, S., Goela, J. Eds. Springer: New York, 2006; pp 37–68.

Murnaghan, F. D. The Compressibility of Media under Extreme Pressures. *Proc. Natl. Acad. Sci. U.S.A.* **1944,** *30*, 244–247.

Nanda, B. R. K.; Dasgupta, I. Electronic structure and magnetism in doped semiconducting half-Heusler compounds. *J. Phys.: Condens. Matter,* **2007,** *17*, 5037.

Nowotny, H; Glatzl, B. Magnetic Properties of Cl_b-Type Mn Base Compounds. *Mh. Chem.* **1952,** *83*, 237.

Park, J. H.; Vescovo, E.; Kim, H. J.; Kwon, C.; Ramesh, R.; Venkatesan, T. Octahedral cation site disorder effects on magnetization in double-perovskite Sr_2FeMoO_6:Sr_2FeMoO_6: Monte Carlo simulation study. *Nature* **1998,** *392*, 794.

Pauling, L. The Nature of the Interatomic Forces in Metals. *Phys. Rev.* **1938,** *54*, 899.

Pénicaud, M.; Siberchicot, B.; Sommers, C. B.; Kübler, J. Calculated electronic band structure and magnetic moments of ferrites. *J. Magn. Magn.Mater.* **1992,** *103*, 212.

Perdew, J. P.; Burke, K.; Ernzerhof, M. Generalized Gradient Approximation Made Simple. *Phys. Rev. Lett.* **1956,** *77*, 3865.

Rai, D. P.; Sandeep; Shankar, A.; Ghimire, M. P.; Khentata, R.; Thapa, R. K. Study of the enhanced electronic and thermoelectric (TE) properties of $Zr_xHf_{1-x-y}Ta_yNiSn$: a first principles study. *RSC Adv.* **2016,** *6*, 13358–13358.

Rai, D. P.; Sandeep; Shankar, A.; Khenata, R.; Reshak, A. H.; Ekuma, C. E.; Thapa, R. K.; San-Huang, K. Electronic, optical, and thermoelectric properties of $Fe_{2+x}VAl$. *AIP Adv.* **2017,** *7*, 045118.

Ramirez, A. P. Theory of colossal magnetoresistance in doped manganites. *J. Phys.: Cond. Mat.* **2017,** *9*, 8171.

Roza, A.; Luaña, V. First-principles calculation of structural, mechanical, magnetic and thermodynamic properties for γ-$M_{23}C_6$(M = Fe, Cr) compounds. *Comput. Phys. Commun.* **2011,** *182*, 1708.

Sbiaa, R.; Meng, H.; Piramanayagam, S. N. Frontispiece: Materials with perpendicular magnetic anisotropy for magnetic random access memory. *Phys. Status Solidi RRL* **2011,** *5* (12), 413.

Schwarz, K. CrO_2 predicted as a half-metallic ferromagnet. *J. Phys. F: Met. Phys.* **1989,** *16*, L211.

Serrate, D.; de Teresa, J. M.; Ibarra, M. R. X-ray absorption spectroscopic study on A_2FeReO_6 double perovskites. *J. Phys. Cond. Mat.* **2007,** *19*, 023201.

Shen, Y.; Zhou, Z. Structural, electronic, and optical properties of ferroelectric $KTa_{1/2}Nb_{1/2}O_3KTa_{1/2}Nb_{1/2}O_3$ solid solutions. *J. Appl. Phys.* **2008,** *103*, 074113.

Singh, J. *Electronic and Optoelectronic properties of semiconductor structures*. Cambridge University Press: Cambridge, UK; 2003, ISBN-I3-978-0-521-82379-I.

Slater, J. C. The Ferromagnetism of Nickel. II. Temperature Effects. *Phys. Rev.* **1936,** *49*, 931.

Snyder, G. J.; Toberer, E. S. Complex thermoelectric materials. *Nat. Mater.* **2008,** *7*, 105.

Soulen Jr, R. J.; Byers, J. M.; Osofsky, M. S.; Nadgorny, B.; Ambrose, T.; Cheng, S. F., et al. Measuring the spin polarization of a metal with a superconducting point contact. *Science* **1998,** *282*, 85.

Strand, J.; Lou, X.; Adelmann, C.; Schultz, B. D.; Isakovic, A. F.; Palmstrøm, C. J.; Crowell, P. A. Electron spin dynamics and hyperfine interactions in $Fe/A_{l0.1}Ga_{0.9}As/GaAs$ spin injection heterostructures. *Phys. Rev.* B **2005,** *72*, 155308.

Tritt, T. M. Thermoelectric Phenomena, Materials, and Applications. *Ann. Rev. Mater. Res.* **2011,** *41*, 433.

Tseng, C. W.; Kuo, C. N.; Lee, H. W.; Chen, K. F.; Huang, R. C.; Wei, C.-M.; Kuo, Y. K.; Lue, C. S. Semi-metallic behavior in Heusler-type Ru2TaAl and thermoelectric performance improved by off-stoichiometry. *Phys. Rev. B* **2017,** ***96***, 125106.

Ullakko, K.; Huang, J. K.; Kantner, C.; O'Handley, R. C.; Kokorin, V. V. Large magnetic-field-induced strains in Ni_2MnGa single crystals. *Appl. Phys. Lett.* **1996,** *69*, 1966.

Umetsu, R. Y; Kobayashi, K; Fujita, A; Kainuma, R; Ishida, K. Magnetic properties stability of $L2_1$ and B_2 phases in the Co_2MnAl Heusler alloy. *J. Appl. Phys.* **2008,** *103*, 07D718.

Urbaniak, M. Spintronic devices–applications. Magnetic materials and nanoelectornics—properties and fabrication. http://www.ifmpan.poznan.pl/~urbaniak/Wyklady2014/index2014.html (lecture no. 6).

Van Dorpe, P.; Van Roy, W.; De Boeck, J.; Borghs, G. Nuclear spin orientation by electrical spin injection in an $Al_xGa_{1-x}As/GaAs$ spin-polarized light-emitting diode. *Phys. Rev.* B **2005,** *72*, 035315.

Waki, S.; Yamaguchi,Y.; Mitsugi, K. Superconductivity of Ni_2NbX (X=Al, Ga and Sn). *J. Phys. Soc. Jpn.* **1985,** *54*, 1673.

Wang, D.; Wang, G. First-principles study the elastic constant, electronic structure and thermoelectric properties of $Zr_{1-x}Hf_xNiPb$ ($x = 0, 0.25, 0.5, 0.75, 1$). *Phys. Lett. A* **2016,** 381, 801–807.

Wang, Z.; Vergniory, M. G.; Kushwaha, S.; Hirschberger, M.; Chulkov, E. V.; Ernst, A.; Ong, N. P. R. J. Cava; Bernevig, B. A. Half-Heusler alloy LiBaBi: A new topological semimetal with five-fold band degeneracy. *Phys. Rev. Lett.* **2016,** *117*, 236401.

Webster, P. J.; Ziebeck, K. R. A. *Alloys and compounds of d-elements with main group elements*; Wijn, H. R. J., Ed.; Springer: Berlin, 2006; Landolt-Börnstein, New Series, Group III vol 19; pp 75–184.

Wikipedia. *https://en.wikipedia.org/wiki/Thermoelectric_generator.*

Wolf, S. A.; Awschalom, D. D; Buhrman, R. A; Daughton, J. M; von Molnár, S; Roukes, M. L.; Chtchelkanova, A. Y.; Treger, D. M. Spintronics: A Spin-Based Electronics Vision for the Future. *Science,* **2001**, *294*, 1488–1495.

Wurmehl, S.; Fecher, G. H.; Kandpal, H. C.; Ksenofontov, V.; Felser, C.; Lin, H. J.; Morais, J. Geometric, electronic, and magnetic structure of Co_2FeSi: Curie temperature and magnetic moment measurements and calculation. *Phys. Rev. B* **2005,** *72*,184434.

Xiao, D.; Yao, Y.; Feng, W.; Wen, J.; Zhu, W.; Chen, X. Q.; Stocks, G. M.; Zhang, Z. Half-Heusler compounds as a new class of three-dimensional topological insulators. *Phys. Rev Lett.* **2005,** 105, 096404.

Zutic, I.; Fabian, J.; Sarma, S. D. Spintronics: Fundamentals and applications. *Rev. Mod. Phys.* **2004,** *76*, 323.

Index

I

L

P

R

S

T

U

Printed and bound by CPI Group (UK) Ltd, Croydon, CR0 4YY
23/10/2024
01777702-0001